PARKS ACROSS AMERICA

Viewing God's Wonders Through a Creationist Lens

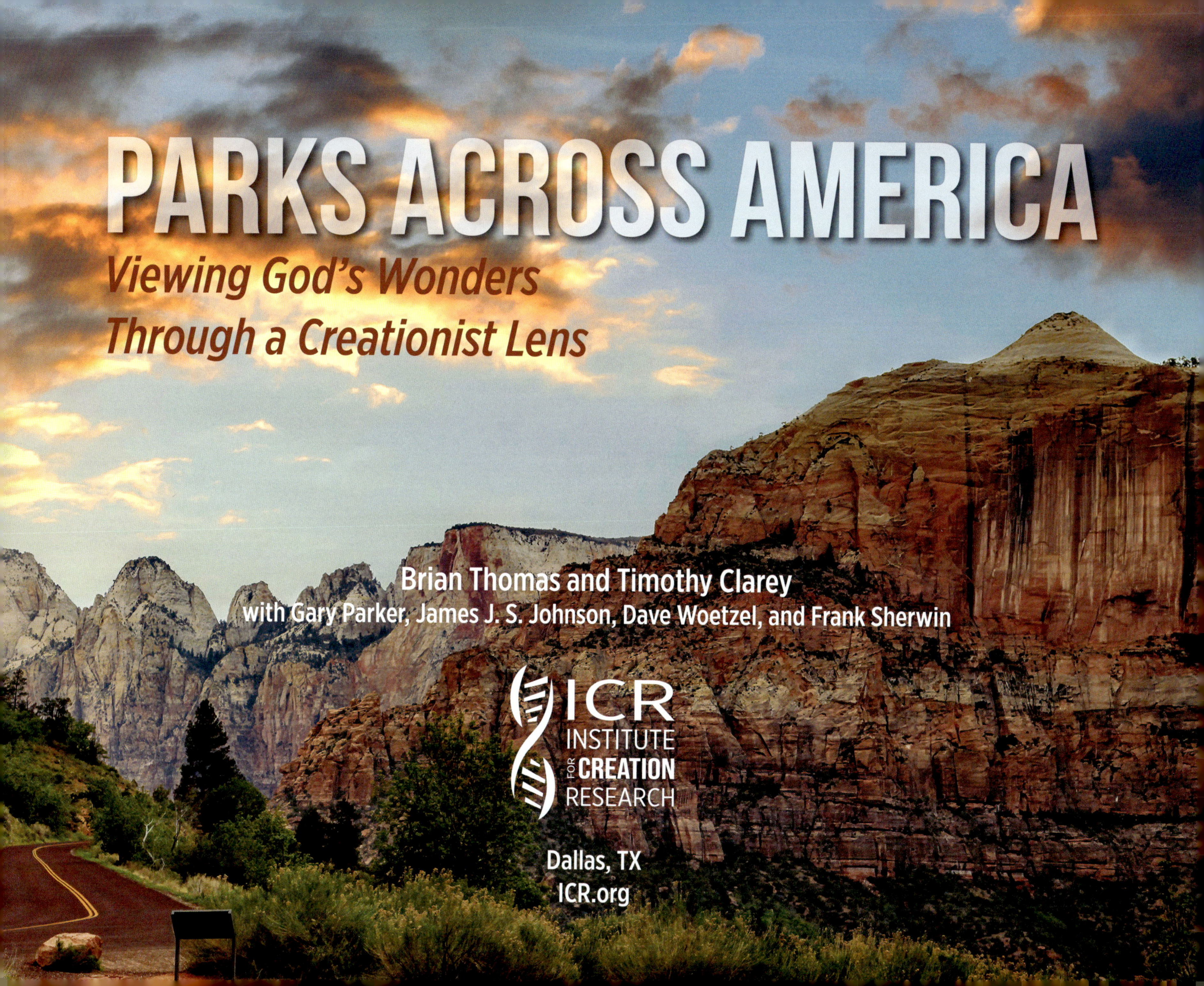

PARKS ACROSS AMERICA

Viewing God's Wonders Through a Creationist Lens

Brian Thomas and Timothy Clarey

with Gary Parker, James J. S. Johnson, Dave Woetzel, and Frank Sherwin

ICR INSTITUTE FOR CREATION RESEARCH

Dallas, TX

ICR.org

PARKS ACROSS AMERICA

Viewing God's Wonders Through a Creationist Lens

by Brian Thomas, Ph.D., and Timothy Clarey, Ph.D., with Gary Parker, Ed.D.,
James J. S. Johnson, J.D., Th.D., Dave Woetzel, M.S., and Frank Sherwin, D.Sc. (Hon.)

First printing: August 2022

All Scripture quotations are from the New King James Version unless otherwise specified.

ISBN: 978-1-946246-98-1
Library of Congress Control Number: 2022941516

Please visit our website for other books and resources: ICR.org

Printed in the United States of America.

TABLE OF CONTENTS

INTRODUCTION

This book is the brainchild of Dr. Brian Thomas. It was his idea to provide a creation-friendly guide to United States parks, both national and state, as well as a few national monuments and lakeshores. This guide is filled with explanations that are refreshingly different from what visitors to these places usually read or hear. The wonderful landscapes are illuminated within the clearly revealed context of God's Word. As a geologist, I feel privileged to have contributed content about many of these parks. Even in the chaos of the Flood, spectacular beauty was created for us to behold. Splendor and design are everywhere in our world.

The 18 chapters of the book originated as regular feature articles in ICR's magazine *Acts & Facts.* They include some of the better-known U.S. parks, including Yellowstone, the world's first national park. Each park showcases exquisite geology and/or biology pointing to our Creator Jesus Christ. Some of the parks reveal strong evidence for the progressive nature of the global Flood recorded in Genesis, while others draw attention to the post-Flood Ice Age. In fact, late-Flood volcanism, like at Yellowstone and Crater Lake, was an essential part of God's plan to create the conditions for the Ice Age.

In addition, many of the animals and plants that fill these parks show clear evidence of design and innate adaptability. After the Flood, animals and plants were able to spread rapidly,

flourish, and fill new environmental niches due to their built-in adaptive capabilities engineered by the Lord Jesus. Organisms have highly regulated innate control mechanisms that enable them to continually track their environments and adapt accordingly. They have adjusted to extreme conditions such as high elevations, deserts, and even caves. Jesus made amazing creatures!

We have included an appendix on Mount St. Helens National Monument because it shows how quickly landscapes can change under catastrophic conditions. The observed rapid deposition and erosion that occurred with the Mount St. Helens eruptions changed the way many conventional scientists view Earth's geological processes. Many of them were forced to acknowledge that major catastrophes have occurred and shaped our world. As creation scientists, we know the global Flood provided the perfect circumstances to explain much of what we see around us. The eruptions at Mount St. Helens are but a reminder of the unfathomable geologic activity during the Flood year. Clear evidence is there below our feet.

As you walk and drive through these parks, I pray this book is a blessing to you. You'll see the same features everyone sees, but your understanding will be guided by true biblical history that reveals many things you won't see on the government signage. And all point us to our Creator, Judge, and Savior, Jesus Christ. We truly have an awesome God who made an awesome world.

Timothy Clarey, Ph.D.
(1 Timothy 1:12)

GRAND CANYON

EXPOSING THE FLOOD

An explorer carefully navigates a canyon wall trail, circa 1930

Grand Canyon carves a 277-mile-long chasm through northwestern Arizona. Running from Lee's Ferry to Lake Mead, the expansive landscape reveals some of the most colorful geology in the world and provides strong evidence for the global Flood.

Lateral Extent of Strata

As you look across Grand Canyon, observe the layers on both walls. The cliffs and the colors match from one side to the other. The gaps between the cliffs were once filled solid, the layers continuous, but the space in between has since been removed by erosion. The bottom flat layers are older and were deposited first; these are called Cambrian system rocks. The youngest layers are on the canyon's rim; these are identified as Permian system rocks.

All of these layers were deposited during the rising phase of the global Flood. Powerful tsunami-like waves spread massive, continuous sedimentary layers for hundreds of miles in all directions across this part of North America. Even relatively thin layers extend across Grand Canyon.

Flat Contacts Show Little Time Between Layers

Grand Canyon's layers are like stacked pancakes. The lowermost flat layer at Grand Canyon is called the Tapeats Sandstone. At about 200 feet thick, it makes a thin, dark brown layer from a distant view. This layer represents the first extensive Flood deposit at this location. The basal boundary of the Tapeats is a special type of unconformity[1] called a nonconformity, where sedimentary rock resides on top of pre-Flood crystalline rock. This surface is also referred to as the Great Unconformity (Figure 1).

Figure 1. *ICR geologist Dr. Tim Clarey addresses a Grand Canyon tour group with the Great Unconformity behind him*

This global phenomenon is found in countless locations where Cambrian system sedimentary layers overlie Precambrian crystalline rocks. Just below the contact, we observe metamorphic crystalline rocks that are oriented nearly vertically, but the overlying Cambrian Tapeats is horizontal. Although conventional geologists claim the Precambrian rock surface here experienced over a billion years of erosion, the contact with the Tapeats is almost perfectly planar—it's flat! Where are the gullies and chasms from billions of years of erosion?

The Redwall Limestone is Grand Canyon's most prominent layer—a red, thick, vertical cliff that spans the middle of the exposed rock layers. It's part of the Mississippian system. Right below it, the Muav Limestone of the Cambrian system appears a bit more grayish. The evolutionary narrative claims 160 million years of erosion occurred between these two rock units. But where are the v-shaped channel patterns that gullies and canyons should have carved on top of the Muav? Instead, one flat limestone lies flat on another nearly everywhere you look.

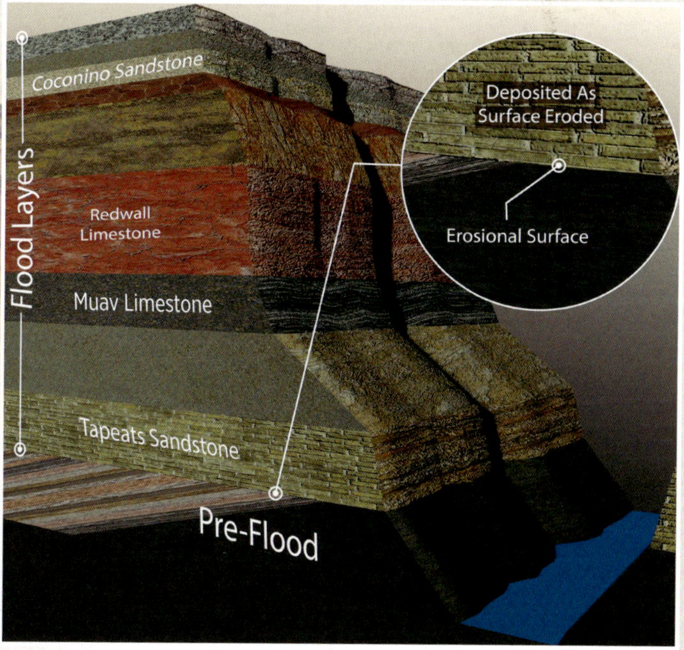

Figure 2. Flat contact shows no erosion and thus no time between the deposition of the Muav Limestone (gray) and the overlying Redwall Limestone (red). The inset shows the Great Unconformity.

Antelope Valley
Hack Canyon
Kanab Creek
Hurricane Cliffs
Uinkaret Plateau
Kanab Plateau
Mount Trumbull
SB Canyon
SB Point
Tuckup Canyon
Tuckup Point
Toroweap Valley
Toroweap Overlook
Parashant Canyon
Whitmore Wash
Colorado River
Supai
Havasu Creek
Hualapai Hilltop
Lake Mead
1221 ft
372m
Grand Wash Cliffs
Grand Canyon National Park boundary
Shivwits Plateau
Andrus Canyon
Prospect Valley
Mohawk Canyon
National Canyon
North
0
10 Kilometers
0
10 Miles
Approximate east-west distance. Map scale varies in this 3D perspective view.
Grand Canyon Skywalk
Sanup Plateau
Burnt Springs Canyon
Granite Park Canyon
Aubrey Cliffs
Surprise Canyon
Horse Flat Canyon
Spencer Canyon
Lower Granite Gorge
Diamond Creek
Peach Springs Canyon

A third flat unconformity lies between the Coconino Sandstone and the Hermit Shale. If you look toward the top of the layers in the canyon, you'll see a thin, light tan-colored cliff on top of a dark red layer. About a million years of erosion supposedly separate these two units (Figure 2). But if they were really deposited millions of years apart, valleys and canyons should be found between each of these layers. Instead, the contact is almost perfectly flat everywhere we see it.

Badlands Topography Shows Catastrophic Carving

By examining the topography of areas that have been catastrophically restructured recently, we can estimate the effects of the Flood and compare that estimation with Grand Canyon's topography. A mudflow following the 1982 eruption of Mount St. Helens created a steep-walled, 140-foot-deep canyon system, complete with side canyons, in a single day.[2] It looks remarkably like a 1:40 scale of Grand Canyon. (See the appendix.)

And we see similar deep gorges and butte-and-basin (badlands) topography associated with the flooding that followed the bursting of Ice Age Lake Missoula. This occurred about 4,000 years ago in the American Northwest. Even conventional geologists now recognize that this catastrophic megaflood created the Channeled Scablands of eastern Washington and widened the Columbia River Gorge.

Today's processes do not form broad, flat rock layers or carve such big and clean canyons. Different processes must be responsible—catastrophic processes. Therefore, it's reasonable to conclude that early floodwaters laid down Grand Canyon rocks while the water was rising, and the receding floodwater carved most of the canyon.

The present narrative for Grand Canyon's formation is insufficient. Most of today's conventional scientists assume the Colorado River slowly carved Grand Canyon over a period of six million years. But today's flow rates simply don't have the power to push the 1,000 cubic miles of rocks and debris all the way out to the Pacific Ocean. Even after millions of years, we should see a lot more talus and debris. Talus refers to rock piles that lean against the cliff walls where they fell. The canyon's base looks like it was swept clean. Only catastrophic water flow and water volume many times greater than what flows there today could sculpt badlands topography and wide gaps in the canyon.[3]

Noah's Flood Explains Grand Canyon

Conventional interpretations still cannot adequately explain why the Colorado River cuts right through the Kaibab Uplift that formed on the western flank of the Colorado Plateau. The Kaibab Uplift has warped an arch of rock about 3,000 feet above the surrounding terrain. Water should have flowed around it, not uphill and through it.[4]

How do Flood geologists explain this? The year-long Flood narrative in Genesis helps us understand what we see. After sediments like the Coconino, Redwall, and Tapeats were deposited in the rising phase of the Flood, the Colorado Plateau was pushed up 5,000 feet during the receding phase of the Flood. This caused the floodwaters to drain off rapidly. Grand Canyon is on the western edge of that plateau. When packed wet sand is lifted up, it stretches and cracks. Water naturally follows the cracks and fractures. Some of those would have run through the Kaibab Uplift to create a path for lots of water to flow downhill and carve a vast canyon.

Rapid uplift and drainage of the receding floodwaters provide both the path-

Havasu Falls, a waterfall of Havasu Creek located in Grand Canyon

way and the necessary volume of water to quickly carve Grand Canyon.[4] This was all accomplished before the Ice Age began. How do we know? Because there are about 150 lava flows that originated on the Uinkaret Plateau and poured down into Grand Canyon during the Ice Age. The canyon had to exist before these lava falls poured in.[5]

The lateral extent of the sedimentary layers, flat contacts between the layers, and vast badlands topography all point to rapid deposition and ultra-high-volume erosion. The rocks reveal the enormity of the global Flood, and the canyon exhibits catastrophic carving from receding floodwaters. Grand Canyon is a reminder of the immense power of the judgment of the Flood.

ARCHES NATIONAL PARK

SCULPTURES FROM THE FLOOD

West of the Rocky Mountains in Utah, Arches National Park preserves some 2,000 natural stone arches. Only a handful of them have the grandeur to attract large crowds, and these are glorious! Countless cameras have captured these vaulted and sculpted sandstones, yet seeing them in person demonstrates why mere photos can't do them justice.

Visitors may ponder how the arches formed. Official park stories leave listeners on shifting sand, but two key observations show how a Flood-based scenario puts arch origin explanations on solid rock.

A Rocky Start to the Arches Story

First, there isn't enough debris to fit the official story. The Arches National Park picture book says:

> Uplift of the entire region fractured the rocks of the Entrada Sandstone. Erosion has worked on these cracks and formed rows of parallel ridges, or fins. The innumerable grains of sand that comprise the fins are held together by a natural cement. This mortar is dissolved by water trickling down the surface of the rock. Sand grains are loosened, and slowly the wall erodes. Water seeping into the porous rock also dissolves cement inside the rock. If these dissolution processes create a hole through the rock fin, an arch is formed.[1]

Some parts of that story make sense, but other parts don't. If en-

Co-author Dr. Brian Thomas stands in South Window Arch at dawn with Turret Arch behind. Two key observations fit a Flood-friendly arch formation scenario.

The sun rises behind Double Arch

Landscape Arch. The scattered rubble beneath the arch fell in 1991 and in a few prior collapse events. Even at today's slow collapse rate, arches could not remain aloft for millions of years.

tire regions uplifted, then why didn't erosion leave behind flat plains over eons instead of these dramatic rocky outcrops and arches? And where do we see erosion carving parallel fins?

Yes, natural cements hold some sand grains together in sandstone, but these can form fast today, such as in creek beds that carry mineral-rich waters. Yes, sandstone erodes slowly by rain, but the surfaces below the arches don't have the enormous piles of sand that eons of today's rain-based erosion would have left behind.

Erosion happens much faster when whole boulders collapse. A few rock piles are visible—perhaps a few thousand years' worth—but after millions of years all the pillars and arches should have collapsed and rubble should completely cover even whole cliffsides.[2]

On the hike to Landscape Arch, a sign explains that some sandstone broke off in 1991. Other rocks fell away in the 1940s. The few scraps of clearly visible rubble from these events still rest below Landscape Arch. The park's most iconic arch, Delicate Arch, has no rubble at its base. Something swept the ground clean.

Again, the park book says, "Slowly the wall erodes."[1] It may happen slowly today, but that doesn't mean it happened slowly in the past. If erosion happened at today's slow rates for eons, then rubble and sand grains should fill the valleys. Where is the rubble from evolution's eons? Perhaps a colossal power washer removed the sedimentary layers that once connected now-separate mountainsides.

Double Arch

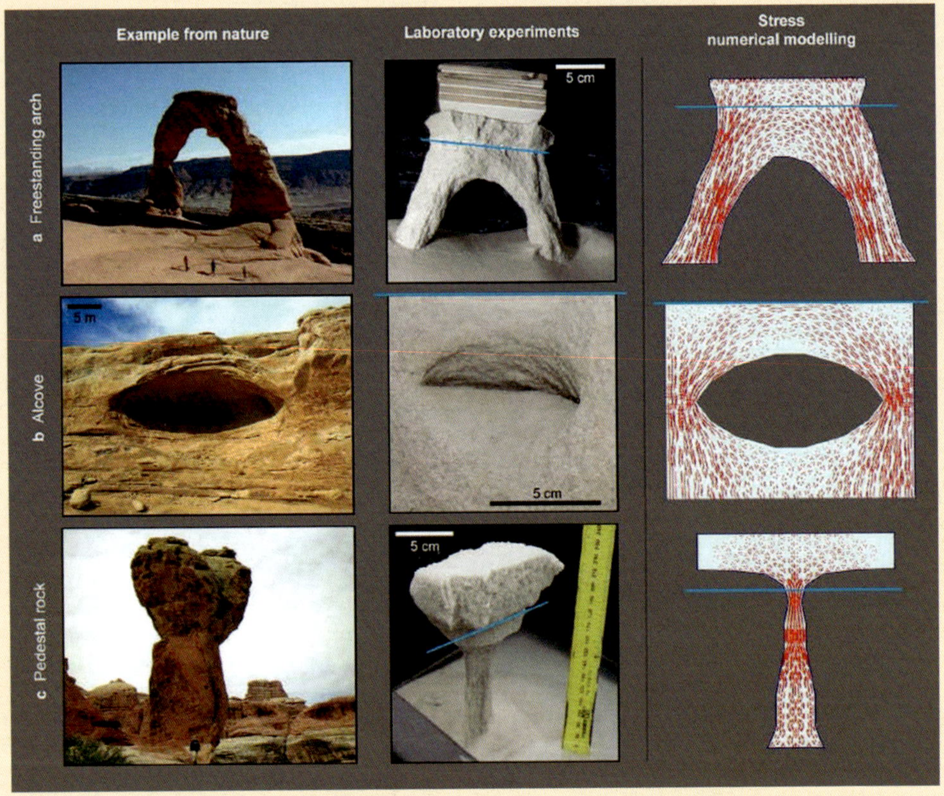

Experiments show that the weight of sand, not cement between sand grains, forms arches in mere hours while sand is partially submerged. The blue lines show water level. Also, the land beneath Delicate Arch, shown top left, represents canyons and cliffsides that were swept clean. Where are the piles of sand and rubble from millions of years of erosion?

Image credit: Reprinted/adapted by permission from Springer Nature: *Nature Geoscience*, Sandstone landforms shaped by negative feedback between stress and erosion, Jiri Bruthans et al, copyright © 2014.

Making Arches in a Lab

A 2014 research study makes up the second key observation. Scientists weighted the top of a block of sand that was wet enough to hold together but not cemented into stone.[3] They raised the water level around the sand block. As sand fell away under different setups, pillars, alcoves, and arches formed fast. These experiments refute the park book's statement that "if these dissolution processes create a hole through the rock fin, an arch is formed."[1]

It happened in minutes when the sand was partially submerged, not over eons of rain. The researchers discovered that sand grains lock together when under pressure. They visited Landscape Arch after their experiments and noticed that the rubble "completely disintegrated" after it fell.[4] Without pressure from overlying rock, the once-hard sandstone rapidly turned to sand. No cement, no rain, and no eons.

Water Draining from Noah's Flood

Gigantic arches and alcoves imply that Utah once lay beneath deep water. The Bible tells us the Flood covered all the earth—that includes Utah. But as the floodwaters lowered, massive crustal movements

Mesas and valleys. The sun sets on a pillared mesa just west of Utah's Arches National Park. During the Flood, water levels lowered while the whole region uplifted from beneath. These receding waters flushed strata toward today's oceans, leaving behind erosional remnants like mesas and arches.

Double O Arch

lifted land upward. This force fractured sandstones into parallel fins. While still partially submerged, the weight of strata above the Entrada Sandstone solidified sand into arches, pillars, and mesas. Descending waters flushed sand and debris toward the seas. When the Flood's water levels dropped enough to start carving channels instead of overtopping the whole landscape, they left behind buttes, valleys, alcoves, and arches as sculptures from the Flood.

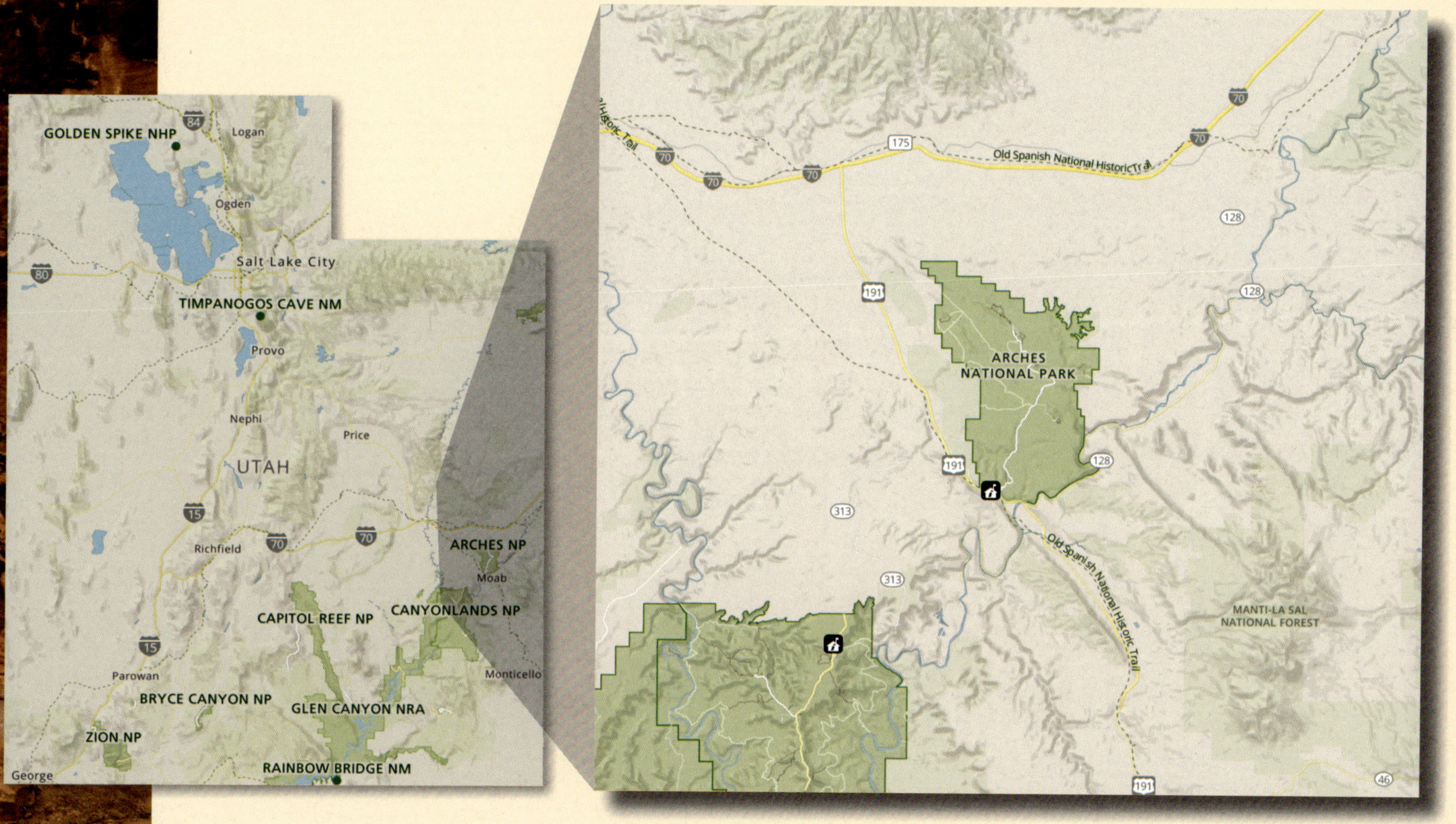

GUNNISON'S BLACK CANYON

THE FLOOD SOLVES MYSTERIOUS MISSING TIME

Figure 1. Black Canyon of the Gunnison National Park's stunning Painted Wall

The Gunnison River winds westward from the Colorado Rocky Mountains through dry and dramatic landscapes. Morning sunlight enlivens Colorado's tallest cliff, the stunning Painted Wall, found in Black Canyon of the Gunnison National Park (Figure 1). Lookout points along the park's south rim road permit people to peer down into the deep canyon, where birds fly far below and the Gunnison River roars in the distant depths (Figure 2).

But Black Canyon does more than drop jaws. It conceals mysteries, including supposedly lost eons of time.

Earth's First Rocks

Painted Wall is 2,250 feet of almost vertical dark metamorphic crystalline basement rock. Gashes of light-colored igneous rock across its face look like fossilized lightning bolts.[1] Official park books and signage try to explain how this happened, but they weave in some wild storylines.

This basement rock formed early in Earth's history. Conventional and biblical perspectives both acknowledge that crustal rock was cooked and squeezed under pressure. Later, molten material mixed with extremely hot water and was injected into cracks and fissures, putting the painted look on the stunning cliffsides. This once-forceful injection implies much higher energy was at work here than we find today.

Figure 2. ICR's Dr. Brian Thomas gazes over the deep Black Canyon gorge

Genesis speaks of "all the fountains of the great deep" bursting upward until waters covered the earth "under the whole heaven" (Genesis 7:11, 19). Catastrophic plate motion during Noah's year-long Flood would have supplied plenty of energy, heat, and water.

"Eons" That Left No Trace

Conventional geologists assign an age of about 1.7 billion years to this basement rock.[2] Biblical geologists point to only thousands of years. Erosion rates lean toward the latter assessment. Various studies suggest that today's slow erosion rates would take a maximum of about 50 million years to completely wash

Earth's continents into the seas.[3]

The park signage itself points out another mystery that a thousands-of-years age assignment would solve. A sign titled "The Great Unconformity" says:

> The thin, light tan layer you see on the canyon rim is Entrada Sandstone, a mere 165 to 170 million years old. The rock immediately below it was formed in the Precambrian period and is about 1¾ billion years old. More than a billion years is absent from the geologic record of Black Canyon!

What would a billion years of erosion do to the top of this basement rock?[4] Erosion carves valleys between mountains, and gullies or canyons between hills. But the top of the basement rock just beneath the Jurassic Entrada Sandstone is perfectly flat![5] It looks like not even one year elapsed between these features.

Similarly, what would a billion years of earthquakes and tectonic motions do to basement rock? They would leave numerous fissures, tilts, and cracks. But the top of the basement rock looks perfectly horizontal and intact to this day!

Noah's Flood Solves the Mystery

Where are the rocks here that represent early Flood deposits like the Cambrian, Ordovician, Silurian, and Mississippian? None exist below the Entrada. Why? A

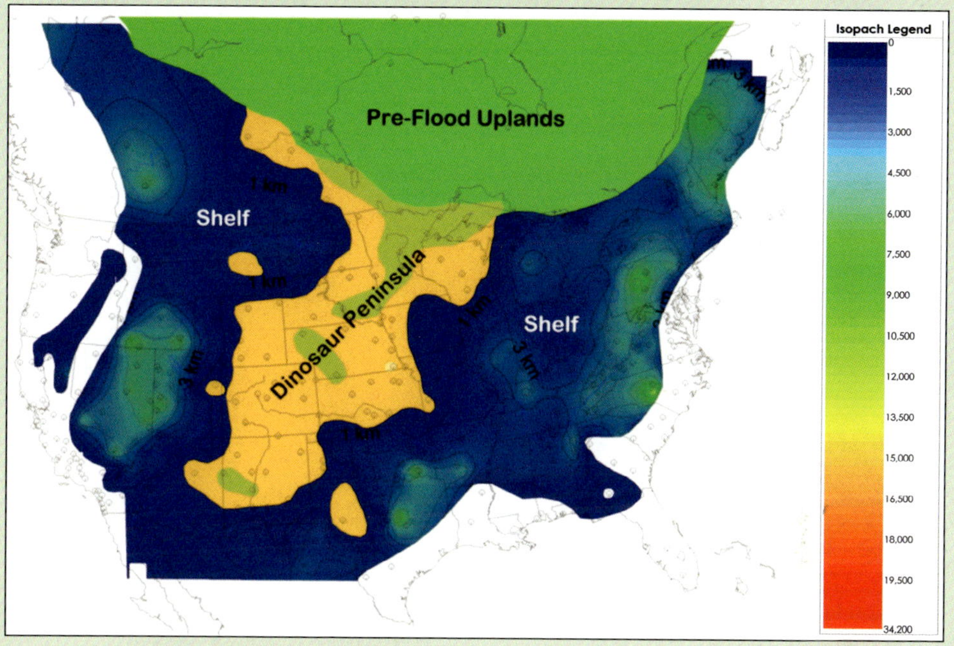

Figure 3. *Thickness map of the early Flood sedimentary rocks, including the Cambrian through Mississippian systems. The measurement is in meters. The Dinosaur Peninsula marked in yellow shows the possible extent of the high ground where dinosaurs were able to escape the early stages of the Flood. The dinosaurs seemed to have fled northward up the peninsula to get away from water rising from the south. Eventually, though, all the land was covered by water, and the dinosaurs perished. Black Canyon is located on the southwestern part of Dinosaur Peninsula.*

view of rock thicknesses across North America reveals an answer (Figure 3).[6] Most likely, these earlier rock systems were never deposited here. They weren't here to erode away!

In a biblical Flood model, areas without early Flood deposits were probably higher in elevation in the pre-Flood world. Pre-Flood lowlands collected early Flood deposits. Since western Colorado didn't accumulate these early layers, floodwater didn't reach this region's heights until later in the Flood year.[7]

The Flood model accounts for Jurassic system rocks lying on the flat crystalline basement rocks with no earlier sediments in between (Figure 4). This area stayed relatively dry during the Flood's first months. Later, tsunami-like waves reached these higher elevations as they flooded the whole earth.

Key Lessons from Black Canyon

Black Canyon exposes one of the wildest stories ever to wind its way into science—a billion years of Earth history that left no trace. These rocks show that the billion years never really happened.

Recent, rapid, high-energy Flood processes better explain the flat contact between basement rock and the Entrada Sandstone. Since the Flood deposited that sand onto the basement rock only thousands of years ago—not 170 million—no wonder it shows no tilts or folds. There hasn't been enough time for tectonics to leave a trace.

The Entrada Sandstone exposed at Black Canyon provides strong evidence that the one-time judgment event described in Genesis solves the missing time mystery.

Figure 4. *Detail of the Great Unconformity, i.e., the basement rock/Entrada contact. One billion years between these strata would have left ruts and tilts. Instead, rapid deposition during a progressive flood fits this flat contact.*

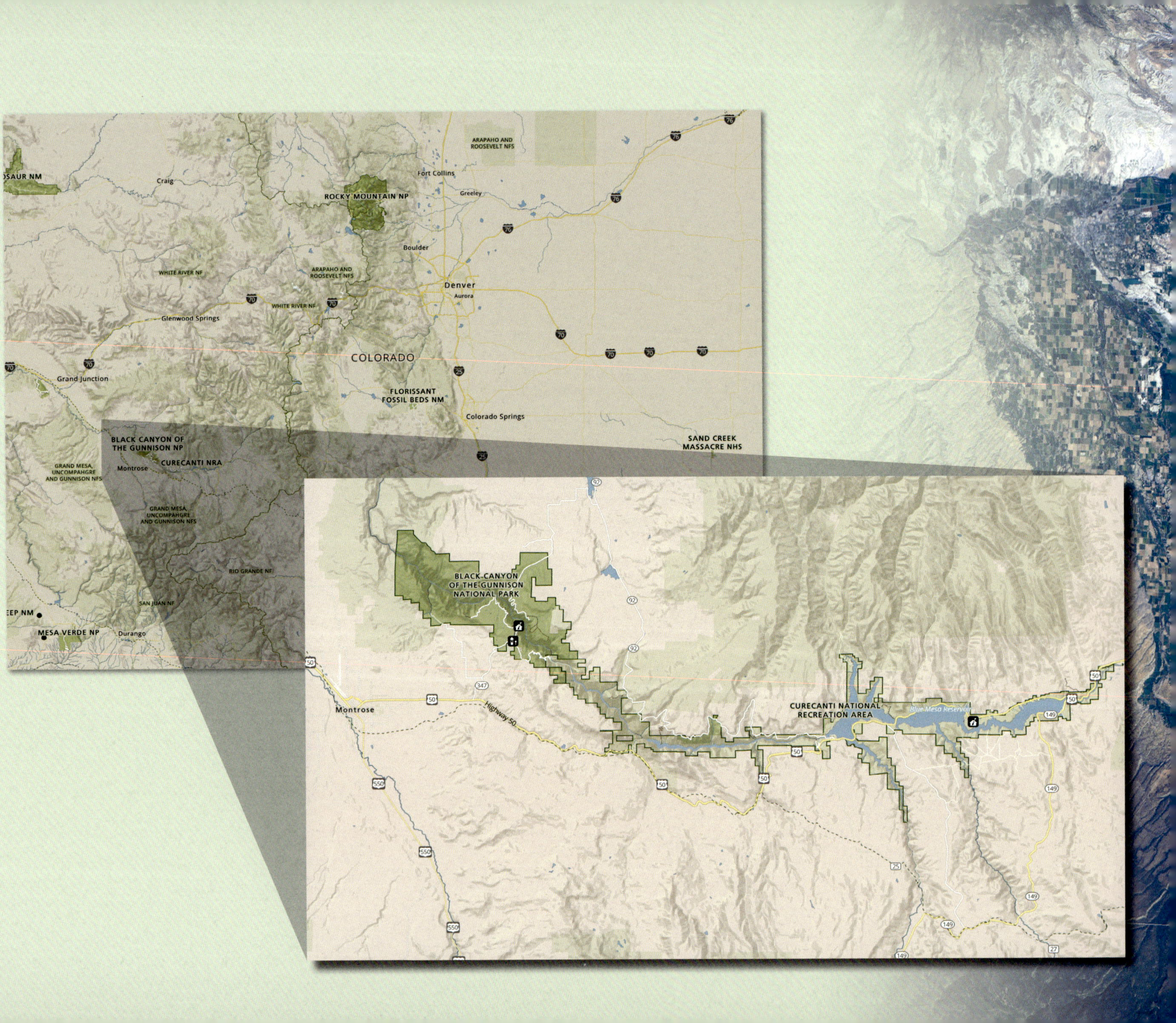

OSAUR NM
Craig
ARAPAHO AND ROOSEVELT NFS
Fort Collins
Greeley
ROCKY MOUNTAIN NP
Boulder
WHITE RIVER NF
ARAPAHO AND ROOSEVELT NFS
Denver
Aurora
WHITE RIVER NF
Glenwood Springs
COLORADO
Grand Junction
FLORISSANT FOSSIL BEDS NM
Colorado Springs
BLACK CANYON OF THE GUNNISON NP
CURECANTI NRA
SAND CREEK MASSACRE NHS
GRAND MESA, UNCOMPAHGRE AND GUNNISON NFS
Montrose
GRAND MESA, UNCOMPAHGRE AND GUNNISON NFS
RIO GRANDE NF
SAN JUAN NF
EEP NM
MESA VERDE NP
Durango
BLACK CANYON OF THE GUNNISON NATIONAL PARK
Montrose
Highway 50
CURECANTI NATIONAL RECREATION AREA
Blue Mesa Reservoir

Satellite view of Black Canyon of the Gunnison National Park

Black Canyon of the Gunnison National Park

THE PAINTED DESERT

FOSSILS IN FLOODED MUD FLATS

Blue Mesa, Petrified Forest National Park

The Painted Desert stretches across 120 miles of northern Arizona. Its sedimentary rocks show bright orange, gold, white, and even purple layers (Figure 1). Amidst this desert lies Petrified Forest National Park, named after its famed stone tree trunks. Here, sands and muds covered both strange and common creatures. Fossils include armored reptiles like aetosaurs, agile dinosaurs like *Revueltosaurus*, and clams.

What happened on the earth to trap these animals and plants in single layers of sediment that sprawl from Idaho to Texas? Some explanations fit these facts better than others.

Fictional Rivers

It's hard to avoid seeing copies of the book *Dawn of the Dinosaurs* in the Painted Desert Visitor Center.[1] Although the Painted Desert receives little rain today, its rock layers paint the picture of a watery start. *Dawn of the Dinosaurs* calls upon fast-flowing water to explain these rocks and fossils.

The Chinle Formation placed in the early phases of the Flood (Upper Triassic) contains the many petrified trees in the park. It extends from Wyoming to Texas and from Oklahoma to Nevada. *Dawn of the Dinosaurs* says this area formed when "energetic, sediment-rich rivers moved across the region."[2] What about burying big amphibians? The book explains that perhaps "a river seasonally flooded over its banks."[3]

But today's seasonal floods don't result in fossils, so why

Figure 1. *The Blue Mesa Member of the early Flood (Triassic) Chinle Formation, exposed in the Tepees at Petrified Forest National Park, includes fossils of the rhino-like* Placerias

A shattered petrified tree near the top of the Blue Mesa

should we believe they did so in the past? We would need massive sheets of water, not little rivers, to make the colorful layers that blanket such enormous regions.

Conventional geologists recognize sheet erosion, a process the book doesn't mention.[4] Enough water can clear everything in its path. At the Painted Desert, fluid has carried and buried sediment and creatures across a vast flat area (Figure 2). In contrast, rivers stay near their channels. Look at how the rocks in the park extend continuously for hundreds of square miles without clear river channels. Do some uniformitarian geologists misidentify sheet flow features as ancient rivers? The same rock layers seen at the Painted

Figure 2. *ICR's Dr. Brian Thomas faces winter winds at the Painted Desert*

Desert and far beyond Arizona demand unimaginably more floodwater and force than regular rivers can deliver.

New Mexico's Ghost Ranch, north of Santa Fe, exposes more of the same sediments. There, dinosaurs and other fossils (Figure 3) pooled in place "after floods."[5] Certain Ghost Ranch quarries contain hundreds of *Coelophysis* dinosaur skeletons that "look as if they were deposited in a single event, probably a massive flood."[6] It's hard to envision the colossal scale and depth of water these deposits imply.

Figure 3. *The Ghost Ranch quarries contain dragon-like* Vancleavea *skeletons like this example that preserves skin scale structure*

Figure 4. *The fossil trees got their color from quartz-rich groundwater percolating through volcanic ash. This process replaced much of the original wood with quartz and agate. Impurities like iron, copper, and carbon made the beautiful mixtures of yellow, red, green, blue, and the black hues.*

Fictional Soils

Typical tales tell of ancient soils that accumulated on river floodplains to explain colorful banded layers visible across the Painted Desert.[3] However, these look nothing like soil horizons we see today. Real soil profiles reveal roots and a gradual increase in organic material toward the top. The Painted Desert's distinct, not gradual, layers lie continuous for miles in all directions.

Petrified Trees from Catastrophes

The Chinle Formation's Blue Mesa Member has world-famous petrified tree trunks up to 180 feet long. The once-low-lying area "was flooded at some point, drowning the trees."[2] But the simple drowning of trees doesn't break trunks off their roots. Petrified trees in the Painted Desert rarely have roots (Figure 4). Silica-rich (quartz) minerals from widespread volcanic activity helped petrify them.

What force could have sheared trees at ground level, ripped off all their limbs, and buried the logs in sediment? Forty years ago, in mere minutes the main Mount St. Helens volcanic explosion stripped thousands of tree trunks of bark, roots, and branches. (See the appendix.) When massive tsunami waves come onshore, they also shear off vegetation en masse. Much bigger volcanic blasts and tsunami waves that occurred during the Flood year help explain the Painted Desert's petrified trees.

Land and Sea Collide

One also expects a worldwide watery catastrophe to mix land and sea creatures, and that's what these rocks show. *Dawn of the Dinosaurs* says, "The Blue Mesa Member preserves a variety of aquatic and terrestrial creatures,"[7] and the same words apply to other formation members. Table 1 shows fossil animals and plants from mixed environments in the Chinle Formation.

Fossils include mammals; reptiles like turtles, lizards, dinosaurs, and crocodilians; frogs; and birds.[8] Shark and coelacanth fossils show up too. Coelacanths live today at about 500 feet or more deep in the sea.[9] Most sharks are also marine. It takes a cataclysm to slam ocean onto land.

Key Lessons from the Painted Desert

ICR's scientists interpret the geologic layers exposed in the Painted Desert as major episodes of the rising waters during Noah's global Flood. The Absaroka Megasequence of rock layers looks similar on all of Earth's continents.[10] During this phase, waters flooded broad pre-Flood lowlands and coastal regions. Tsunami-like waves carried in marine animals like coelacanths and sharks and mixed them with wetland animals like crocodilians. Every continent has land and marine fossils buried together—powerful evidence for the global Flood.[11,12]

River systems and soils today look nothing like the Painted Desert's rocks and fossils. The enormity of the Flood explains the wide spread of each rock layer. Volcanic eruptions during the Flood contributed ash and silica to petrify violently buried trees. Draining floodwaters carved the landforms. Since the Flood explains the Painted Desert's main features, this park in Arizona paints a picture of the reliability of Genesis.

General Creature Type	Specific Creature Kind	Creature's Setting
Dicynodonts	*Placerias*	land
Amphibians	Frogs	water + land
Birds	*Protoavis*	air + land
Lizards	*Trilophosaurus*	land
Reptiles	Aetosaurs (var. sp.)	land
	Dromomeron	land
Crocodilomorphs	*Pseudosuchus*	water + land
	Phytosaurs (var. sp.)	water + land
	Revueltosaurus	water + land
	Vancleavea	water
Arthropods	Mite (plant galls)	land
	Bark beetle (bore holes)	land
Crustaceans	Clam shrimp	water
	Crayfish	wetland
Chondrithyans	Shark teeth	water
	Coelacanth	water
Osteichthyans	Fish scales, bones	water
Bivalves	Unionoid clams	water
Gastropods	Snails	water
Dinosaurs	*Coelophysis*	land
	Chindesaurus	land
Mammals	Treeshrews	land
Conifers	*Agathoxylon*	land
Cycads (Sago palms)	*Sammiguelia*	wetland
	Charmargia	wetland
Lycopods	(var. sp.)	wetland
Tree ferns	*Itopisdema*	wetland

Table 1. *Plant and animal fossils in the Chinle Formation and its equivalents*

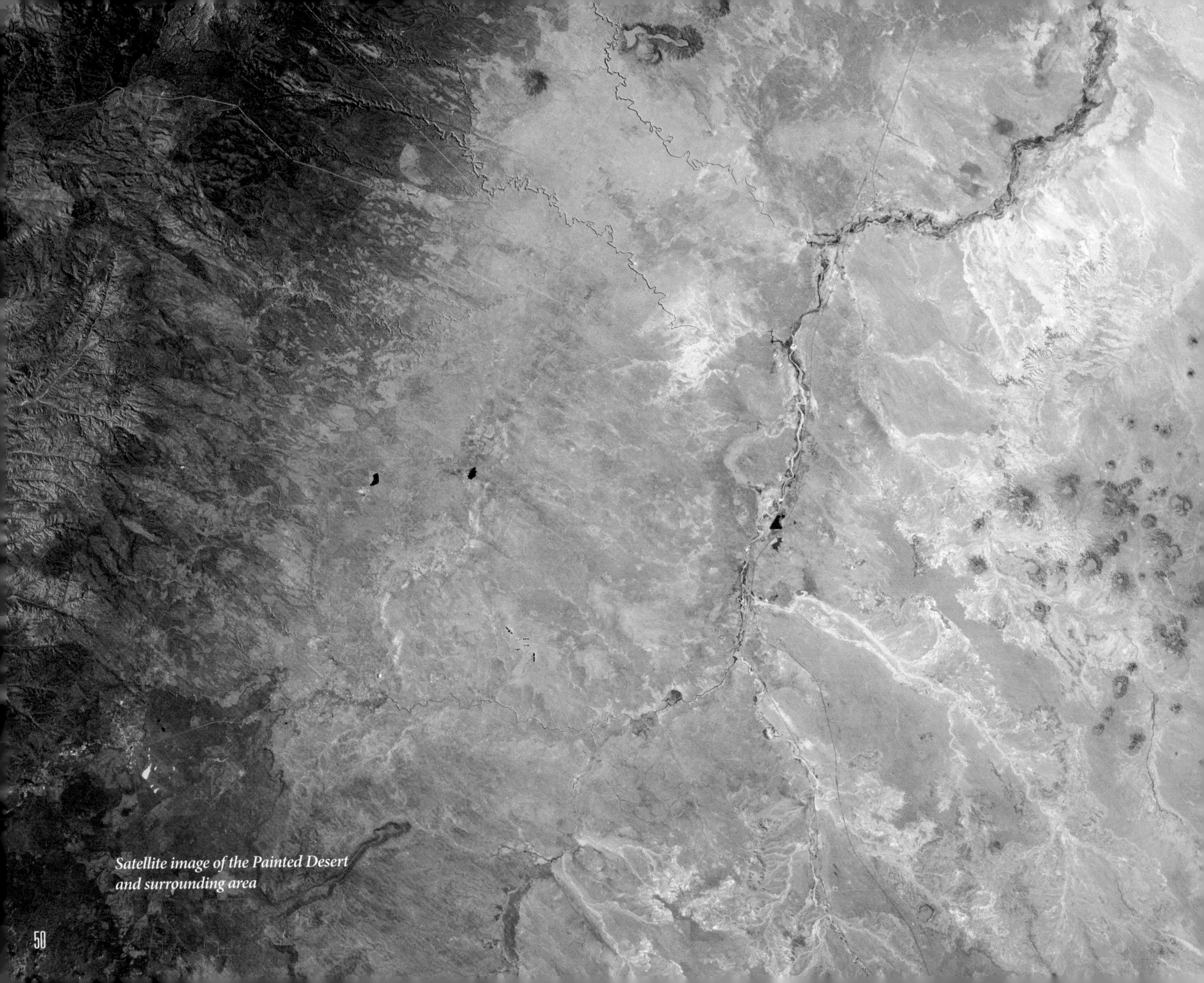

Satellite image of the Painted Desert and surrounding area

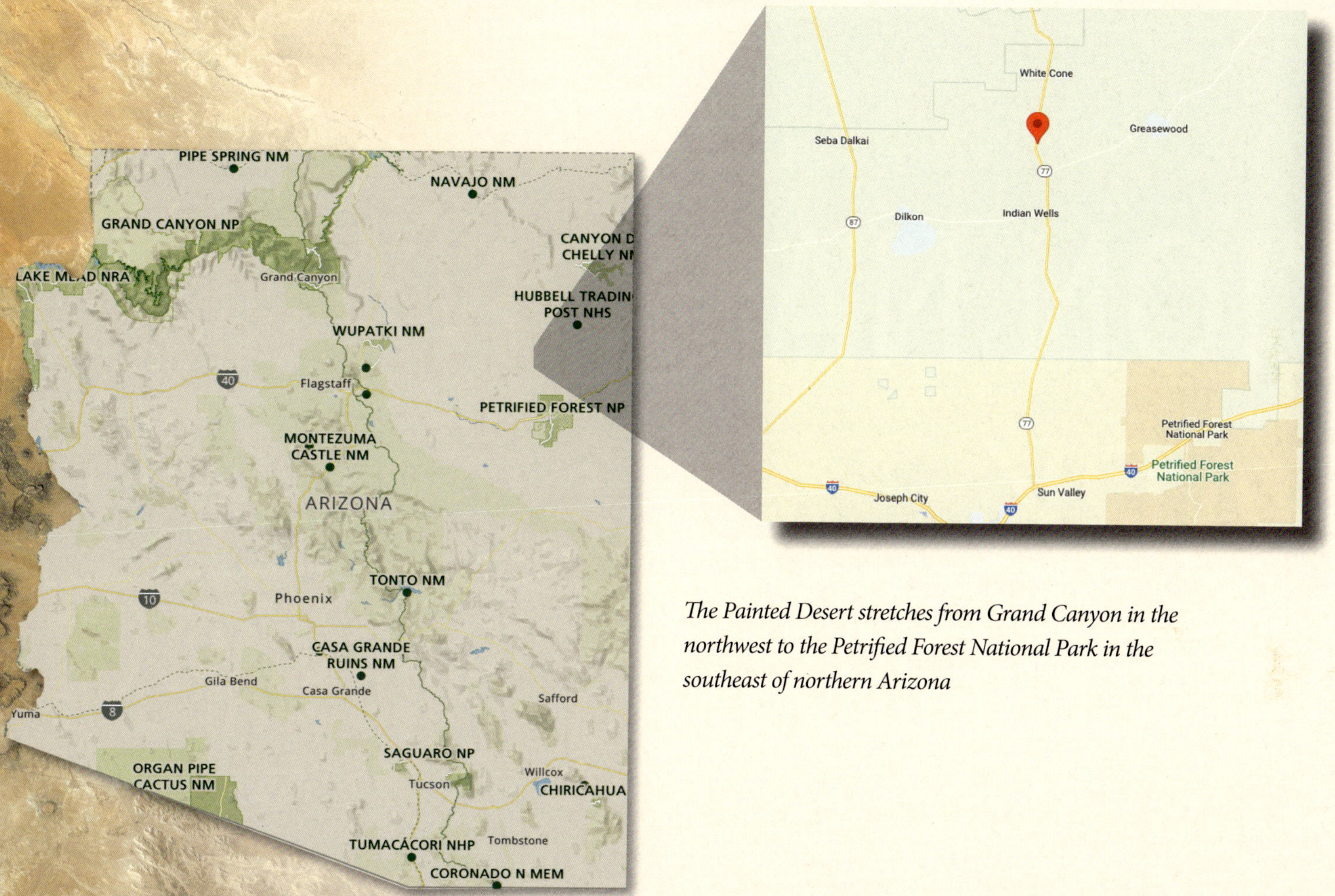

The Painted Desert stretches from Grand Canyon in the northwest to the Petrified Forest National Park in the southeast of northern Arizona

ZION NATIONAL PARK

EVIDENCE OF DEEP WATER SAND WAVES

It was like driving into another world when we emerged from Zion-Mount Carmel Tunnel right at dawn. Zion National Park opened up to us in brilliant colors, and our jaws dropped at the splendor of sunlit sandstone cliffs. We joined the procession that picked its way down the switchbacks as passengers begged their vehicles' drivers to slow down for photos of the spectacular valley walls.

How many of the millions of annual visitors wonder how and when these awesome rocks arrived?

Cross-Bedded Sandstones

Zion National Park exposes some of the largest and clearest cross-bedded patterns among Earth's immense and extensive sandstone layers. In the past, continuous flow piled sand dunes on top of one another, preserving the internal structures that cut at an angle across the flat bedding surfaces. Cross-bed is the technical term for this angle. The U.S. Geological Survey says about Zion's dunes that "the cross-bedding found in the Navajo Sandstone is the result of sand being moved by wind."[1] This anticipates a vast desert. Of course, wind does build sand dunes on land, but few know that sand dunes also form underwater. Could the sandstone layers at Zion be from water?

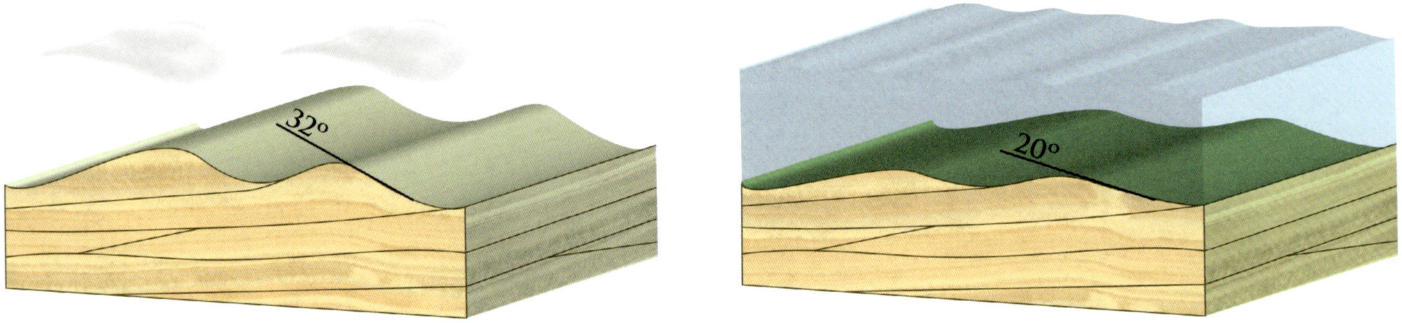

Figure 1. *Experiments show that wind deposits sand dune slopes at angles of about 32°, whereas water deposits sand dune slopes closer to 20°. Wind and water flow direction is left to right.*

Figure 2. Zion National Park's Checkerboard Mesa

Scientists experiment with wind and water to make sand dunes in the lab. Their findings reveal a straightforward method to discern between dunes caused by wind and dunes caused by water. Figure 1 shows the key difference between these two sediment transport modes.

Generally, cross-beds form when wind or water transports sand up the gentle windward side of a dune. Sand drops as it reaches the slower flow on the leeward (downwind) side, creating a steeper slope in front. Even after the sand becomes sandstone, the sharp angles (cross-beds) against the flatter bedding below are preserved. Cross-beds formed by air are steeper than the ones formed by water.

Cross-beds from windblown sand form near the stable slope angle of dry sand at between 32° and 35° from horizontal. In contrast, angles not much greater than 20° indicate that fast-flowing water put the sand there. So, the angle measure of a cross-bed layer reveals whether wind or water formed the original dune. Figure 2 shows cross-beds from stacked dunes in a mesa at Zion. Figure 3 shows them in black and white for ease of analysis.

Measurements at Zion's Checkerboard Mesa were between 17° and 16°—the angle water would have caused. The Navajo Sandstone even has limestone and dolomite mixed with some of the sandstone layers.[2] These carbonate minerals form in water. The area was certainly covered in water. So much for dogmas about desert dunes in Zion.

Figure 3. Sandstone cross-beds at Zion National Park lie at angles consistent with water deposition despite consensus assertions of wind deposition

What Difference Does This Make?

If wind built these dunes long ago, then the narrative would have to change. The water-deposited rock formations found above and below

The road through Zion National Park at sunset

them in adjacent regions suggest it took eons for deserts to replace oceans and that it took more eons for oceans to cover deserts. But empirical evidence of water-deposited rocks all the way up flushes the eons far away.

Hikers walking in the Virgin River through Zion Narrows

Deep, Deep Water

Experiments with dunes have taught us about more than just angle measurement. They've taught us that if the dune is bigger, the water must be deeper. Sand dune heights are equal to about one-fifth of the water depth that forms them.[3] Some dunes in the Navajo Sandstone reach 95 feet high,[4] meaning they formed beneath about 500 feet of fast-flowing water! The tsunamis that made these sand waves must have been gigantic—and larger than any observed in recent history.

Science, Scripture, and Salvation

When we take the history of Scripture at face value, we can model some expectations of how Noah's Flood transformed the landscape of the world. Then we can compare these expectations with what we find in the real world. When we see what destructive and devastating scars the Flood cut into the earth, we behold the judgment of God. As 2 Peter 3:6 says, "...by which the world that then existed perished, being flooded with water." Here, Peter warns that since God has judged the whole world with water once already, we'd better prepare for the judgment to come. Massive, water-formed dunes bolster Peter's warning.

The Flood model uses data from the effects of tsunamis and amplifies the patterns of destruction. Models of waves larger than today's tsunamis show them ripping up rocks and depositing layer upon layer of sediments for months during the Flood year.[5] Water deposition in a rising flood would leave stacked sandstone layers that match the ones we observe in Zion—layers stacked flat with no erosional channels filled in.

Other stratigraphic data suggest the sandstones at Zion National Park were deposited while the floodwaters were approaching their zenith on a global scale.[6] Ocean crust formed rapidly during the Flood to help push water higher and higher, generating larger and larger tsunami waves.[6] This was the perfect recipe to stack thick cross-bedded sands on top of each other. The Flood model shows that the rock stack exposed at Zion formed in a few weeks during the Flood year, and experiments confirm that the model fits.

In this way, science increases confidence that the Bible's history makes the clearest sense of our world. No wonder it also makes the most sense of spiritual realities like our need for rescue from our sins through the Lord Jesus Christ. He paid our penalty and offers everlasting life to all who repent of sin and trust in Him. Reading the clues from underwater dunes at Zion National Park increases the urgency of the message in 2 Peter 3.

Zion National Park offers dramatic beauty

Cross-bedded sandstone formed by deep water waves

GOLDEN SPIKE NHP
84
Logan
Ogden
Salt Lake City
80
TIMPANOGOS CAVE NM
Provo
Nephi
Price
UTAH
15
Richfield
70
70
ARCHES NP
Moab
CAPITOL REEF NP
CANYONLANDS NP
15
Monticello
Parowan
BRYCE CANYON NP
GLEN CANYON NRA
ZION NP
RAINBOW BRIDGE NM
George

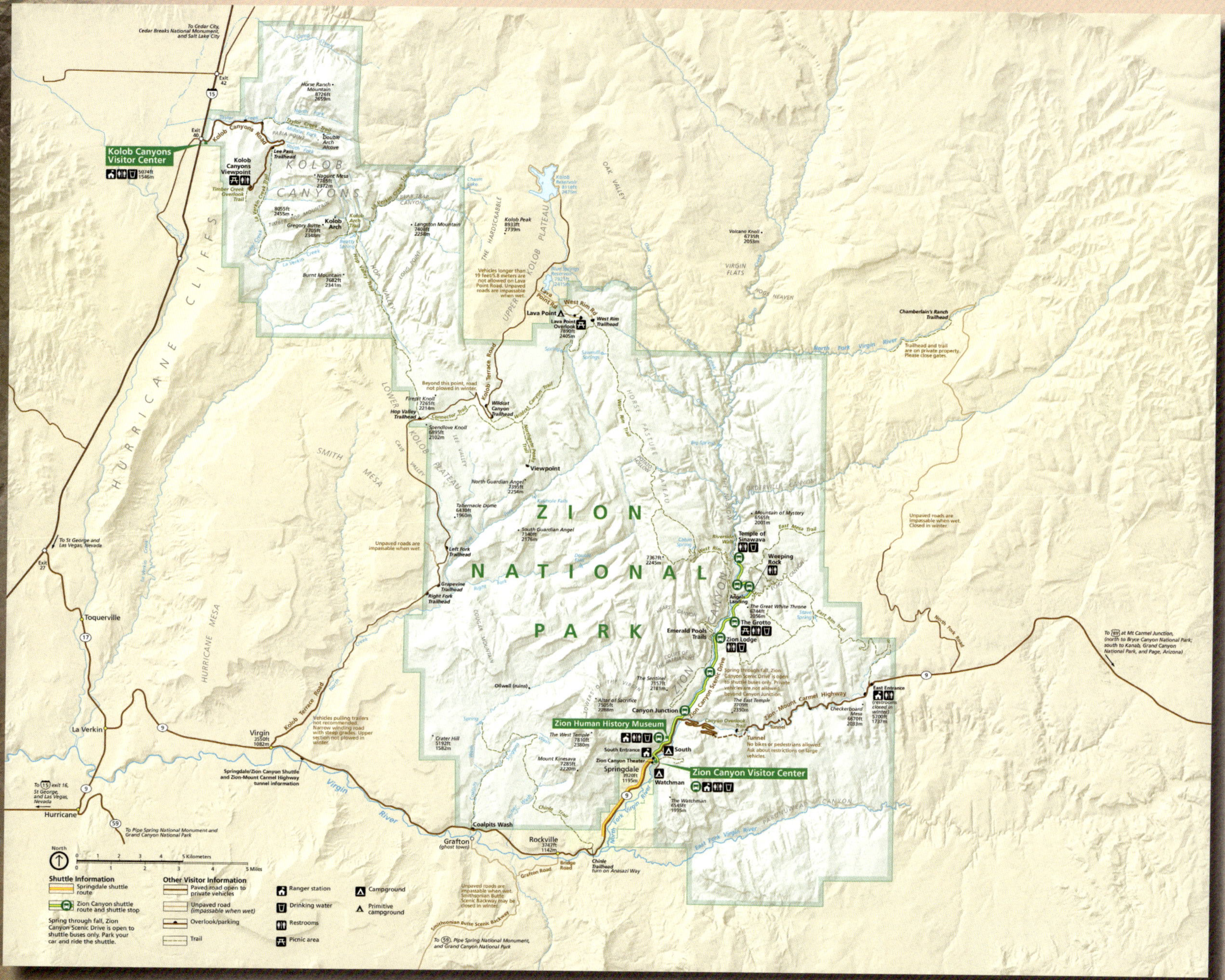

ZION NATIONAL PARK
Kolob Canyons Visitor Center
KOLOB CANYONS
Kolob Canyons Viewpoint
Kolob Arch
HURRICANE CLIFFS
Lava Point
West Rim Trailhead
Kolob Terrace Road
Hop Valley Trailhead
Wildcat Canyon Trailhead
Viewpoint
Left Fork Trailhead
Grapevine Trailhead
Right Fork Trailhead
Temple of Sinawava
Weeping Rock
The Grotto
Zion Lodge
Emerald Pools Trails
Canyon Junction
Zion Human History Museum
South Entrance
South
Watchman
Zion Canyon Visitor Center
Springdale
Zion-Mount Carmel Highway
East Entrance
Tunnel
Virgin
Rockville
Grafton (ghost town)
Coalpits Wash
Toquerville
La Verkin
Hurricane
Virgin River
HURRICANE MESA
SMITH MESA
Spring through fall, Zion Canyon Scenic Drive is open to shuttle buses only. Park your car and ride the shuttle.
Shuttle Information
Springdale shuttle route
Zion Canyon shuttle route and shuttle stop
Other Visitor Information
Paved road open to private vehicles
Unpaved road (impassable when wet)
Overlook/parking
Trail
Ranger station
Drinking water
Restrooms
Picnic area
Campground
Primitive campground
North
5 Kilometers
5 Miles

THE EVERGLADES

TWO BIOLOGY BASICS THE BIBLE GOT RIGHT

Imagine a river 50 miles wide and 100 miles long but only inches deep. Its slow flow is hidden under tall swamp grasses punctuated by small, rounded clumps of cypress. Home to gators and snakes, the portion of the Everglades preserved as Everglades National Park in south Florida is worth visiting during the day, but it's perilous to pitch your tent there at night. The Everglades house animal interactions that illustrate two biology basics the Bible got right.

Kinds Make Kinds

As God commanded in Genesis 1, living kinds generate more of their own kinds. Each kind has its variations while sticking to its basic body design. For example, flowers maintain their essential identities even though their petals may display different colors.

Similarly, cats exhibit various coat patterns and body sizes, but they remain cats. Even cats with different body and coat types have the ability to interbreed in a continuum. For example, lions can cross with tigers and pumas with ocelots.

In the Everglades, the Florida panther (*Puma concolor couguar*) prowls. But not without a struggle—the cat has faced controversy. Back in the 1990s, Florida media cried doom for the big cat. I, Dr. Gary Parker, wrote, "The small population was riddled with so many mutations affecting its circulatory and reproductive systems that mating adults could not produce a cub that could survive even one year."[1]

I told my college classroom that wildlife biologists could turn this around if they brought in some big cats from the West. This would reintroduce genetic health into the Florida panther population. Then, I took them on a field trip. We walked up in time to hear a park ranger explain that to restore genetic vigor, workers were crossbreeding the Florida panther with panthers (cougars) from out West. Thirty heads swiveled around to look at me.

In that moment, those students understood that the Bible got its creature kinds right. Genesis 1:24 says, "Then God said, 'Let the earth bring forth the living creature according to its kind.'" Cats, whether they are small or large, whether they look striped, spotted, or smooth, always produce cats.

Similarly, pythons produce pythons. After Hurricane Andrew toppled Miami pet stores in 1992, Burmese pythons (*Python bivittatus*) got loose in the Everglades. Their population has since grown. Scientists recently discovered crossbreeds between Burmese and Indian pythons (*Python molurus*). *Reptiles* magazine said these variants "may further diversify the gene pool of the large constricting snakes in south Florida and could perhaps change the population dynamics of the species and the localities in which they could migrate to."[2] In other words, it's possible that the hybrids have the genetic gear to help them pioneer new territory—maybe they'll slide into your backyard!

Invasive python in the Everglades

Human intervention and natural means both can produce crossbreeds. But why is it that crossbreeds boost genetic strength? From a biblical worldview, the answer is straightforward. Those variants descended from separated populations of the same created kind.[3] Regathering long-separated genes masks the harmful effects of mutations that had been accumulating.

Violence and Death Intrude

When driving through Alligator Alley on our way to the Keys, we almost always see alligators. Like algae-covered logs bobbing along the stale waters, the creatures look serene from a distance. But they live an eat-or-be-eaten lifestyle. Biologists and rangers alike have found those gators eating pythons, and pythons eating them. One official park photo, shown on the next page, found fame for its horror. A python apparently swallowed an alligator that proved much too big. The gator's bulky carcass ruptured the python from within, leaving both creatures dead. We certainly no longer live in that "very good" paradise God created in the beginning (Genesis 1:31).

If you frowned in reaction to the image captioned "Reptile Remorse," then your face betrayed the fact that you sense something is very wrong with this world. Death does not belong in God's creation. If it did, we would pay no mind to such gore. Rather, death rudely intrudes into our world. The second biological basic the Bible got right—and we see it in the Everglades in spades—is that violence violates the original good God created.

The book of Romans confirms the origin of death given in Genesis, saying, "Through one man sin entered the world, and death through sin, and thus death spread to all men, because all sinned" (Romans 5:12). Not just humans, but "the whole creation groans" with pain from sin's deathly result (Romans 8:22). How can we ever escape this place of death? We too deserve the judgment of God. His wrath is revealed against our ungodliness and our suppression of the truth. But praise God, the Lord Jesus took our death penalty and rose from the grave in order to rescue us! After He returns, He promises to build a new world where "there shall be no more death, nor sorrow" (Revelation 21:4).

The Everglades show what Genesis 1–3 describes: kinds make kinds, and death intrudes. The Bible's trustworthy biology secures our confidence in Scripture's promise that the Lord Jesus is our way to God.

Reptile Rancor. Alligator (Alligator mississippiensis) *takes advantage of a python* (Python bivittatus) *in Everglades National Park.*

Reptile Remorse. This alligator proved too big for its python predator to process. The alligator ruptured the python's body, leaving both animals to rot in the Everglades.

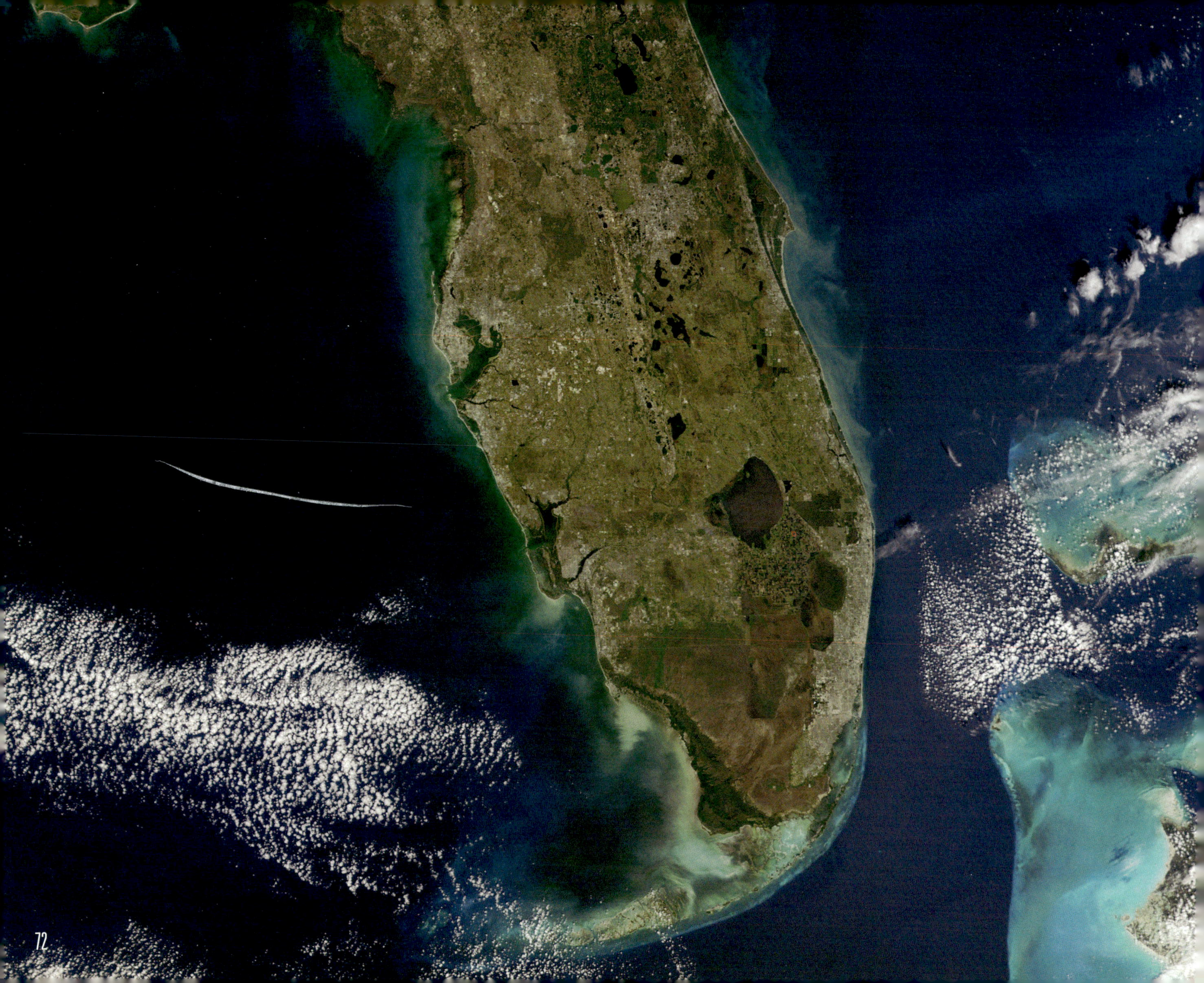

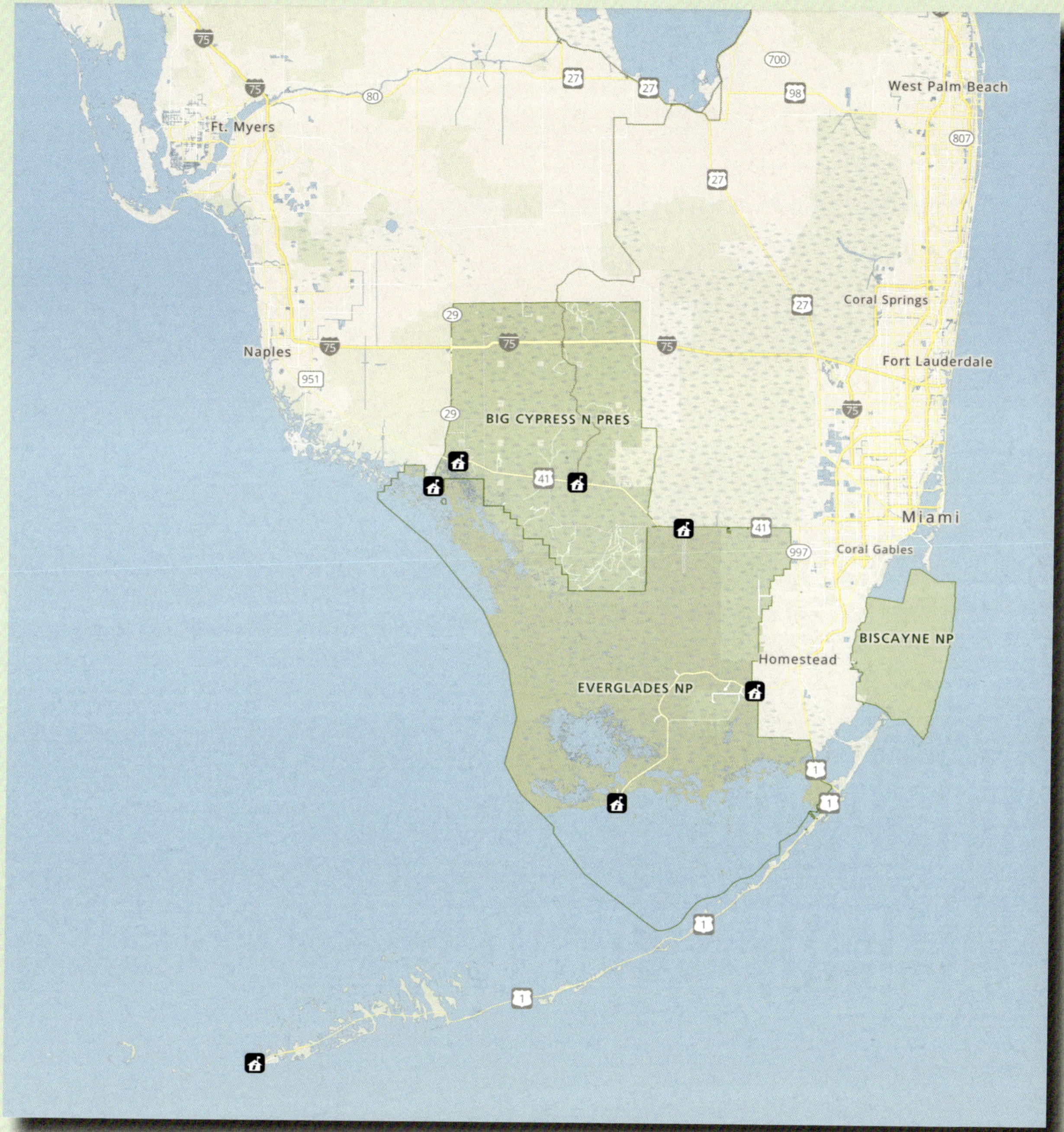
West Palm Beach
Ft. Myers
Coral Springs
Naples
Fort Lauderdale
BIG CYPRESS N PRES
Miami
Coral Gables
BISCAYNE NP
Homestead
EVERGLADES NP

ACADIA

THE MAINE JEWEL AMONG AMERICA'S PARKS

Acadia National Park displays Christ's artistry throughout almost 50,000 acres of forests, island wildlife, salty-air breezes, and rocky seashores. When visiting Acadia, we can see exhibits of how Christ loves life, loves variety, and caringly fits wildlife for habitats.

What Is Acadia National Park?

Composed of Maine's peninsular mainland and nearby islands in the Atlantic Ocean, Acadia is America's first national park established east of the Mississippi River.[1] Rising upon Desert Island is Acadia's Cadillac Mountain, the tallest mountain (1,530 feet) on America's eastern seacoast, blending granite boulders, evergreen trees, and cranberry bushes.[1,2]

Acadia's visitors can enjoy appreciating creation ecology—especially wildlife and biogeography—in beautiful eastern forests, freshwater ponds and streams, marshy wetlands, biodiversity-packed islands, and seashore galore.[1,2]

Busting Deep-Time Myths of Ancient Granites

Cadillac Mountain showcases pink-colored, coarse-grained granites that are well worth hiking. These granites are early Flood igneous rocks exposed by erosion—including post-Flood Ice Age glacial erosion—that formed rounded, smoothed surfaces, especially on Cadillac Mountain.

Evolutionists claim magmas move and cool slowly over eons of time. But recent research debunks this uniformitarianism-based myth, proving that granite magma formations and movements don't require millions (or even thousands) of years. Accordingly, Acadia's granites corroborate Genesis.[3,4]

God Bioengineered Life to Be Fruitful and Multiply

Wildlife abounds in Acadia. Evergreen-dominated forests house birds, butterflies, bugs, large and small mammals, reptiles, amphibians, earthworms, and more.[2] Freshwater habitats (streams, ponds, marshlands) contain busy aquatic insects, finfish, shellfish, turtles, frogs, etc.[5] Tidewater-washed shorelines support shorebirds, seabirds, crustaceans, hauled-out seals, jellyfish, and more.[6]

Christ programmed Acadia's wildlife for fruitfulness and multiplication,[7] so Acadia's wildlife abundance continues generation after generation. But none of this is simple or lucky. Christ selected (and bioengineered) the necessary details for the reproductive success of Acadia's wildlife. Documenting those details could fill many libraries, but space constraints permit only one example here.

Consider the eastern tiger swallowtail (*Papilio glaucus*) but-

terfly. It begins life as a large, green, fertilized egg deposited upon a plant, afterward hatching into a crawling larval caterpillar, and then transforming into an immobile cocoon before later emerging as a flight-capable butterfly. All of this sequential development constitutes a truly miraculous metamorphosis.[8]

Should we be surprised that God's program for procreating and producing fragile butterflies is carefully regulated, successfully reproducing swallowtail butterflies? Also, should we be surprised that God incorporated camouflage traits (including what look like snake eyes, nose, and mouth) to offset the vulnerabilities of growing swallowtails?[8] Furthermore, as caterpillars they use foul-smelling emissions to deter predators![9]

Not only does God accomplish reproductive success with such butterflies, He does so efficiently, cleverly, and beautifully. Acadia's wildlife neighborhoods provide many more examples of habitat-filling life cycles—including growth, development, and reproduction "according to their kinds"—and ICR has already described many of these in other publications (lobsters, seals, echinoderms, etc.).[10-12]

God Loves Variety in Wildlife

The diversity of Acadia's habitat-filling wildlife is mind-stretching.[1,3] Acadia's forests consist of evergreens (like balsam fir, pines, spruces) mixed with deciduous trees (like maple, birch, aspen), sometimes edged by marshy wetlands or cranberry bogs, generously hydrated by freshwater streams and ponds—plus a hodgepodge of shrub-sprinkled rocky fields and hills that stretch to Acadia's rocky coasts, cobblestone shores, mudflats, tidepools, and sandy beaches.

Acadia's forests contain birds (owls, hawks, woodpeckers, chickadees, woodcocks, waxwings, nuthatches, doves, spruce grouse, pheasant, sparrows, crossbills, finches, pine siskin), land mammals (moose, fox, deer, squirrels, pine marten, mice, raccoon, bats, snowshoe hare), insects (honeybees, blackflies, deerflies, borer beetles, mosquitoes, yellow jackets, butterflies, moths), and arachnids (spiders, ticks).[2]

Freshwater streams (Marshall Brook, Hunter Brook) and ponds are plentiful in Acadia—Schoodic Beaver Pond, Aunt Betty Pond, Half-Moon Pond, Round Pond, Seal Cove Pond, Witch-Hole Pond, Eagle Lake, etc.—where eagles, ospreys, river otters, and minks snatch brook trout (*Salvelinus fontinalis*). Besides the almost-ubiquitous brook trout, these freshwaters contain brown trout, lake trout, perch, and landlocked Atlantic salmon.

Adjacent wetlands (e.g., Gilmore Marsh, Bass Harbor Marsh) support muskrats, frogs, toads, salamanders, painted turtles, birds (bobolinks, ducks), and garter snakes.[5] Acadia's intertidal shores, tidepools, and coastal seawaters (including Somes Sound fjord) provide habitat for many birds (eiders, ducks, gulls, loons, osprey), marine mammals (porpoise, seal), finfish (sunfish, herring, bass, haddock, mackerel, salmon, cod), shellfish

Rounded knobs of granite smoothed by the Ice Age glaciers that covered Acadia

(lobster, green crab, mussels, sea snails, tortoiseshell limpet), sea anemone, sea stars, etc.[6]

God Makes Wildlife and Habitats Fit Each Other

Much could be said about how Christ fits diverse wildlife to fill Acadia's habitats.[13] Acadia's pregnant pine marten mothers experience delayed embryo implantation, timing gestation so that birthing occurs in springtime when food availability is optimal. Because the bioenergetic metabolism needs (i.e., food energy requirements) of lactating marten mothers are so demanding, both mothers and newborns benefit from this phenological providence.[14]

Whether we see it or not, Christ carefully provides for even His humblest creatures.

Snowfleas are springtail hexapods, insect-like detritivores (eating decaying leaves, moss, fungi) that survive Acadia's cold winters using antifreeze![15]

Acadia National Park—what an adventure-inviting blend of post-Flood mountain rocks, wildlife-rich forests, hiking trails, and tide-washed island coastlines. Acadia uniquely exhibits Christ's caring providences, reminding us that our Creator loves to fit His variety of creatures to fill a variety of habitats.[16]

Maine's coastland shows the influence of southward-moving glaciers that carved the landscape

Ellsworth
ACADIA
NATIONAL PARK
Mount Desert
Island
Ironbound
Island
Long Island
Bartlett
Island
Swans Island
Marshall
Island
Long Island

CRATER OF DIAMONDS STATE PARK

AND THE ORIGIN OF DIAMONDS

At Crater of Diamonds State Park in western Arkansas, families dig diamonds for fun while more serious sifters seek sensational paydays. Countless brides have wondered where the pretty diamonds on their rings came from. A visit to Crater of Diamonds reminds us of two key research results that refute the old ages assigned to diamonds. These results favor the Bible's record of a more recent creation and global flood.

Uncut diamond

An Ancient Volcano

Diamonds adorn the Arkansas state flag and license plates in honor of Crater of Diamonds State Park. The park is a rare place for the public to mine diamonds. The experience draws a diverse crowd. Serious searchers bring wagons toting buckets of gravelly soil to the park's water troughs for wet sifting, while children bring their fingers to poke around the ground.

People have sifted for diamonds at Crater of Diamonds State Park in Murfreesboro, Arkansas, since 1906. To date, more than 70,000 diamonds have been found in the park.

Despite the crater in the name, first impressions reveal no distinct crater shape at ground level. Rather, volcanic rocks called lamproite mixed in the soil show that you're walking on the mouth of an ancient volcano. Park signage compares the ancient volcanic blast cloud to one of the largest ever recorded—the 1848 explosion of Krakatoa in Indonesia.

In a moment, mantle material from miles below shot up through Earth's crust. When it neared the surface, exploding steam blasted fresh lava rock into a vast

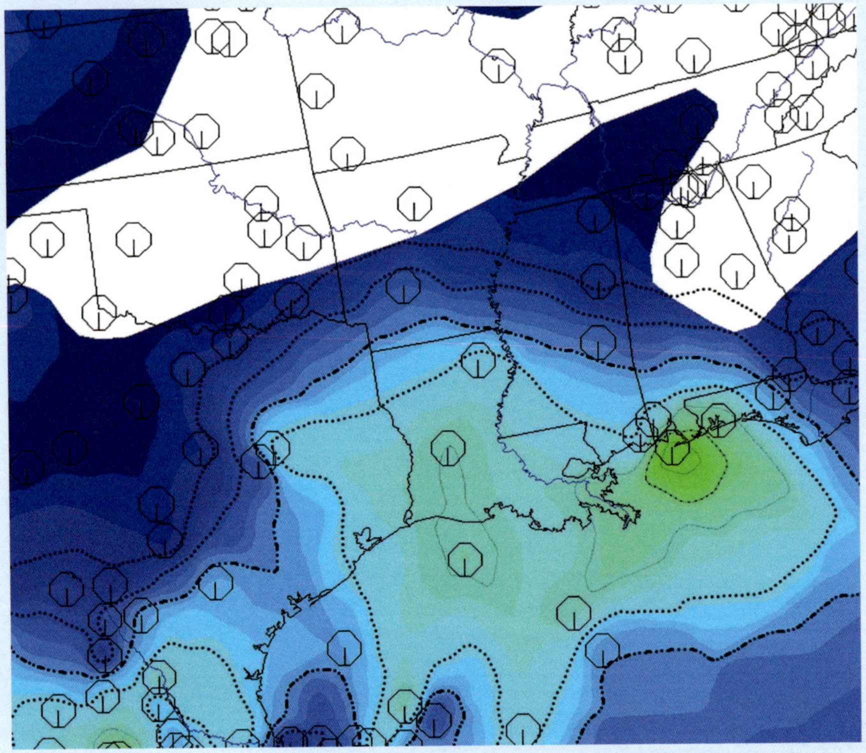

Figure 1. Isopach (thickness) map of the sedimentary rocks composing the Zuni Megasequence (near the high point of the Flood) for Arkansas and surrounding states. The volcano at Crater of Diamonds State Park erupted near the top edge of these sediments, shown in blue. Greener color indicates thicker deposits, and bluer color indicates thinner deposits. Circles show drill core data collection areas.

volcanic ash plume. If this erupted while Cretaceous sediments had already dropped to the bottom of the Flood's surging waters, then it happened near the middle of the Flood year.[1] Figure 1 shows the thickness of sediments that landed on southern Arkansas during the fifth of six major pulses of Flood sedimentation, each pulse called a megasequence. If our Flood model is accurate, then the ash plume could have erupted right into floodwaters, not directly into air.

Back then, volcanoes gushed as continents collided. During this fifth megasequence, months into the Flood year, water finally overtopped the highest hills that then existed (Genesis 7:19). As volcanic eruption rates slowed after the Flood, so did diamond deliveries from the deep. Those who appreciate their diamond rings can thank God for bringing up gems through judgment.

Two Research Results

Does evidence from Earth—or, even better, directly from diamonds—confirm our Flood model?

Various park signs repeat the common assertion that diamonds formed billions of years ago. But two key research results bury that idea. The first came from the ICR RATE project.[2] One branch of RATE research included measuring radiocarbon (radioactive carbon) in diamonds. If Earth formed thousand of years ago, then Earth's diamonds might still retain some radioactive carbon atoms. In contrast,

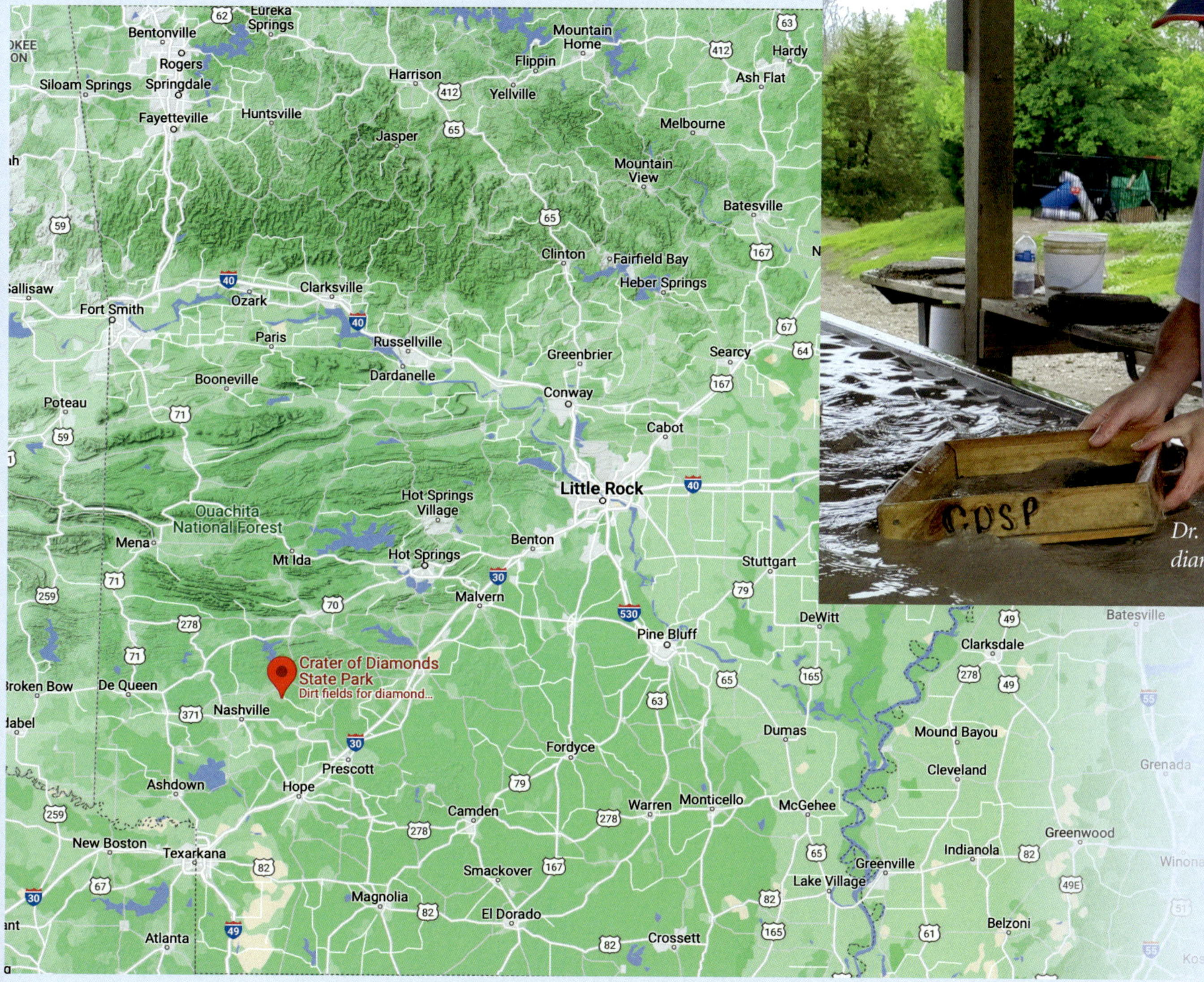

Dr. Brian Thomas wet-sifts soil in search of diamonds at Crater of Diamonds State Park.

a billions-of-years burden bars carbon-containing material deemed older than 100,000 theoretical years from having any radiocarbon.[3]

Radiocarbon labs use carbonaceous earth materials like natural gas, coal, marble, and graphite as instrument blanks. Workers assume they are too old to have any radiocarbon, but labs consistently reveal more radiocarbon in the blanks than contamination can reasonably account for. And before the RATE project, nobody had tested diamonds.

We express the levels of radiocarbon using pMC. This refers to the percentage of radiocarbon found in an ancient sample compared to the radiocarbon content in a standard modern sample. In general, the older the material, the smaller the percentage. Based on the measured decay rate, samples older than 100,000 theoretical years should show zero percent radiocarbon after subtracting known contamination sources like radiocarbon atoms in the lab air.

In 2005, Dr. John Baumgardner described the RATE team's stunning results, shown here in Table 1.[4] All 12 African diamond specimens had radiocarbon in them! With all this carbon still radioactive, how could these diamonds be even one million years old, let alone billions? No known underground process could generate the measured levels of radiocarbon.

Sample ID	pMC	±
NMBclr1	0.31	0.02
NMBclr2	0.17	0.02
NMBclr3	0.13	0.03
NMByel1	0.09	0.02
Letlhakane-3	0.07	0.02
NMBrn2	0.07	0.02
Letlhakane-1	0.04	0.03
NMByel2	0.04	0.02
Orapa-F	0.03	0.03
Kankan	0.03	0.03
Kimberley-1	0.02	0.03
Orapa-A	0.01	0.03
AMS sensitivity	0.002	

Table 1. Twelve RATE-tested diamonds showed levels of radiocarbon above the AMS (accelerator mass spectrometer) theoretical sensitivity limit.[4] It appears that short-lived radiocarbon is intrinsic to these diamonds. pMC is percent modern carbon, uncalibrated.

Lab ID	pmC	±
UCIAMS-12677	0.018	0.001
UCIAMS-12678	0.017	0.002
UCIAMS-12676	0.017	0.001
UCIAMS-12679	0.016	0.001
UCIAMS-12674	0.015	0.001
UCIAMS-12675	0.015	0.001
UCIAMS-15445	0.021	0.003
UCIAMS-15444	0.018	0.002
UCIAMS-15443	0.015	0.002
UCIAMS-15446	0.013	0.002
UCIAMS-15447	0.011	0.002
UCIAMS-9638	0.008	0.001
UCIAMS-9640	0.006	0.001
UCIAMS-9639	0.005	0.001

Table 2. Six of nine diamonds show levels of radiocarbon far above the instrument's detection limit. The top six measurements were taken from one diamond. UCIAMS refers to the University of California Irvine Accelerator Mass Spectrometer. Results from Taylor and Southon, 2007.[5]

Diamond ring

Volcanic tuff with lamproite from the Crater of Diamonds State Park

Conventional experts, including the late R. E. Taylor, decided to measure their own diamonds. They found radiocarbon in theirs, too. Table 2 summarizes their results, published in 2007.[5]

These kinds of results so irked R. E. Taylor that he wrote a 10-page paper to try and explain them away.[6] He and his co-authors argued that since diamonds are billions of years old, any radiocarbon they measured must result from contamination. That's just circular reasoning.[7] These results deserve to be investigated, not buried beneath bias.

A visit to Crater of Diamonds State Park, and for that matter a look at any diamond ring, reminds us of short-lived radiocarbon that two research projects found inside diamonds. Both sets of results confirm the thousands of years of Earth history that the Bible has been reporting to us all along.

WACO MAMMOTH

NATIONAL MONUMENT

Hunters in the 1970s discovered mammoth bones exposed in sediments near the confluence of the Bosque and Brazos Rivers in north central Texas. They notified scientists, who began excavating the site. Decades later, in 2015, the site and surrounding land became a national monument.

Today, visitors hike a short path to a large, air-conditioned building that covers the dig site. A herd of huge Columbian mammoths got buried in mud. How and when did they die? Waco Mammoth National Monument offers an opportunity to allow God's Word to help answer these questions and to place the Ice Age in world history.

Burying Mammoths

All this mud, now dry ground, demands an immense catastrophe. So much sediment rapidly covered the ancient landscape that it entombed whole mammoths, an Ice Age cat, a giant tortoise, and camel remains.

Park signage agrees. Scavengers and microbes normally recycle dead animal bodies. One sign says, "So why didn't the bones here get recycled? The sed-

iment!" This site received much higher rates of sediment than today's slow processes produce.

Another sign nearer the burial site says, "Scientists think the mammoths may have been gathered at a watering hole when a flash flood struck, trapping the animals in thick mud, killing them." Possibly, huge Ice Age storms dropped so much rain so fast that the runoff made a muddy torrent that filled the shallow valley.

The extent of sediment offers a key clue to connecting the mammoths' flash flood to the Ice Age. Rock layers from Noah's Flood tend to extend very flat across hundreds of square miles. That calls for sheets of water to cover whole continents. In contrast, smaller Ice Age sediment volumes fill lowlands and valleys.

A second key clue to connecting the sediments at Waco Mammoth National Monument to the Ice Age is their position. These valley-confined sediments lie atop widely spread rock layers far below the surface. The Flood's continent-covering waters explain the lower widespread layers. The Ice Age's later storms explain the mammoths' sediments at the surface.

The Ice Age

Noah's Flood best explains the Ice Age.[1] The Flood happened first. It produced untold volumes of hot, molten material to create all-new seafloors. It warmed the oceans. Evaporation increased. Water vapor rose into the sky until it cooled and

Mammoth Q
(Male)

fell back down as rain in Texas' lands and as frozen forms farther north. As residents in Florida can attest (particularly those who watch weather forecasts in September), warmer sea surfaces fuel larger hurricanes. During the Ice Age, Flood-warmed seas powered hurricanes so huge that we call them "hypercanes."[2]

Possibly an Ice Age hypercane gorged itself on warm Gulf waters before dumping its contents onto central Texas. Camels, cats, tortoises, and mammoths caught in low areas had too little time to escape the torrents.

Timing

The National Park Service study on the site says that researchers in the mid-1980s obtained a carbon date of 28,000 years before present (BP) for a mammoth bone. It says this "became the de facto date of the mammoth event."[3] Later researchers tried uranium-series (U-series) dating on mammoth tooth enamel. But these results "were not initially considered accurate because they were much older than the expected age of the site." Why not base ages on dating results instead of expectations?

Later, optically stimulated luminescence (OSL) tests offered an age estimate of 68,000 BP. So, carbon dating, U-series dating, and OSL dating clash.[4] Probably none are correct. Unlike ever-changing and subjective conventional age-dating estimates, the Bible's timeline comes from reliable eyewitnesses.

Mammoths, animals, and mankind migrated from the Ark's landing site in the Middle East. They multiplied into Asia and the Americas. Back then, when much of the world's waters were frozen into ice sheets, land bridges connected Siberia to Alaska.

Using just the Bible and allowing for a few unrevealed details, the Flood happened in 2472 BC (4422 BP), give or take a few decades.[5] All the earth's broad and extensive flat-lying rock layers fell into place during that year. Waco's mammoths must have been buried some centuries afterward.

Mammoth remains at Waco tell a sad tale of Ice Age catastrophe. They don't present a defensible challenge to the history in God's Word. Instead, they highlight how helpful the Bible is in explaining our earth's rocks and fossils.

Camel

JOHN PENNEKAMP CORAL REEF STATE PARK

I, Gary Parker, surely enjoyed leading college student groups on spring break camping trips to John Pennekamp Coral Reef State Park on Key Largo in Florida. Diving excursions to the coral reefs about three miles offshore were my favorite parts, but there were also fascinating boardwalk and canoe trips to the mangrove forests along the shore. Near-shore swimming gives opportunities to meet intriguing sea creatures up close and personal, plus trails wind through exotic forests that grow nowhere else in America. On a late-night walk along a sandy beach, you could stir up flashing points of light by disturbing bioluminescent sea creatures—"fireflies" of the sea. Want to come along?

Tropical reefs like those in the Florida Keys feature corals on shallow sea floors. Each tiny coral animal (polyp) is essentially a tiny stomach with "fingers" (tentacles) for stuffing itself with food. It has no brain, eyes, heart, lungs, or liver and lives in a secreted limestone cup (corallum) cemented to hundreds or thousands of its closest neighbors to form a coral colony. The colony may be shaped like a boulder, brain, antler, fan, or bush that sways in the waves.

Many corals need very little food because their polyps and stony cups house green algal cells that can make their own food by photosynthesis. The algae share their food with the corals, while the coral supplies the algae a safe living space. The relationship between a coral polyp and its algal cells is called symbiosis—specifically mutualistic symbiosis, since both partners benefit. The coral-algae partnership is a relationship "made in heaven" by God's decree at creation!

God also demonstrates His artistic range by produc-

Queen Angelfish

ing a stunning array of colors, forms, and sizes, all expressing distinctive variations on His creative themes. He did the same in creating human beings with the ability to appreciate all that creativity. Colorful, swaying forms of large sea fans, sea plumes, and sea whips are favorites of snorkelers, divers, and visitors aboard glass-bottom boats or in aquariums, who find them particularly beautiful.

The Lord of creation engineers astonishing actions as well as beauty. For example, at fish-cleaning stations, small fish near distinctive corals are willing to clean the teeth of big fish while they literally wait in line! When its turn comes, the big fish swims up, opens its mouth wide, and lets a cleaner fish swim in and pick bits of debris from its teeth. Then the second miracle occurs. The big fish with clean teeth lets the cleaner fish out—before it goes off to hunt other small fish to eat! What effect do you think Darwin's "war of nature" would have on cleaning symbiosis? Cleaning stations showcase fish behaviors in another example of mutualistic symbiosis—and another example of the Lord's handiwork.

The outer barrier reefs set the stage for a dramatic series of ecological—not evolutionary—changes called succession. In now-falsified Darwinian thinking, environments make changes in organisms. In the reality that we call creation science, organisms track and even alter their environments in predictable

Christ of the Deep, located about six miles off the coast of Key Largo

ecological succession. One group prepares an environment for organisms living in a different area to move in. Together they "multiply and fill" the earth (Genesis 1).

Visitors at Pennekamp can trace four stages in ecological succession, beginning with (1) the outer reefs. These provide storm protection for (2) the sandy-bottom grass beds with their patch reefs, a shallow water habitat rich in varied life forms that delight snorkelers and swimmers. Toward shore, (3) mangroves anchor the shoreline with storm-resistant roots, provide shelter for egg deposits, and protect newly hatched larvae from predators. Park visitors use canoes, kayaks, or boardwalks to explore these "nurseries of the sea." Lastly, (4) the walking trails at Pennekamp weave through orchids, lianas, and exotic trees like mahogany in America's premier tropical forest.

Most of Pennekamp's dive sites and ecological zones feature God's handiwork, but one was shaped by a human sculptor's hands. A huge underwater statue called Christ of the Deep stands in 25 feet of water. The statue reminds us that Christ stands ready to save those "lost at sea" as well those lost on land. Whoever turns to Jesus for forgiveness of sins finds a place of peace and joy forever with Him!

Bioluminescent plankton

CARLSBAD CAVERNS NATIONAL PARK

FAST FORMATIONS

Have you ever toured a cave? There is something especially intriguing about exploring the deep, dark underground and its mysterious animals and rock formations. Tour guides at Carlsbad Caverns and at caves around the world repeat the mantra that it takes a thousand years of slow drips to build just a centimeter of these formations, called speleothems.[1] Two observations about speleothems show they can form much faster than that. These observations help explain why we would expect caves and their speleothems to have formed within the Bible's timeline of only several thousand years.

Cave Formations and the Bible

Most modern speleothems typically grow very slowly. Conventional geologists extrapolate today's slow rates into the past to estimate millions of years for some of the large speleothems. Such conclusions depend on whether or not we believe that today's processes (such as rainfall and erosion) happened at the same rates in the distant past. And we know they did not.

Instead, catastrophic activities that follow from the Bible's record of Noah's Flood and the later Ice Age shaped the large geological structures on our planet. And science sings the same tune.

The Ice Age and Caves

The first observation that suggests speedy speleothem formation starts by noting that the Carlsbad region of New Mexico has been dry for centuries. This arid climate means most of

Dripstone formations are called speleothems and take two basic forms. Stalagmites point up from the floor like traffic cones (Figure 1). Stalactites, the most common, hang from the ceiling like icicles. Speleothems form when subtle chemical changes cause dissolved minerals to come out of water (precipitate) and form a solid. Caves are termed dead when they have no active speleothem growth. This happens when the water that had been dripping from above dries up. Thus, climate profoundly impacts speleothem formation rates.

Figure 1

its speleothems are dormant. They therefore largely formed in the past. But when and how fast? Scientifically, we cannot assert that they grew slowly back then if their growth rate is zero today, and nobody was there to measure the growth rates in the distant past. What if Carlsbad's speleothems formed fast in a wetter past?

In general, cave formation likely began with the help of receding waters from the global Flood and the rapid circulation of water underground. Then during the Ice Age, the American Southwest was wet for centuries. Enough water flowed there to quickly deposit Carlsbad's speleothems.[2] And the Ice Age saw wetter climates around the world. In other words, the Ice Age offers a stronger reason to suppose fast speleothems than slow ones.

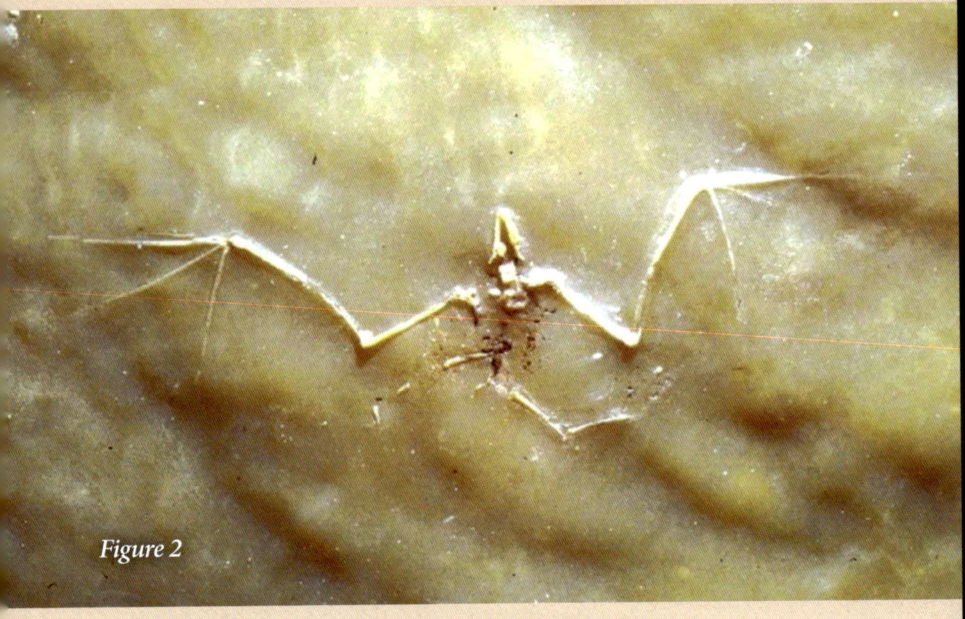

Figure 2

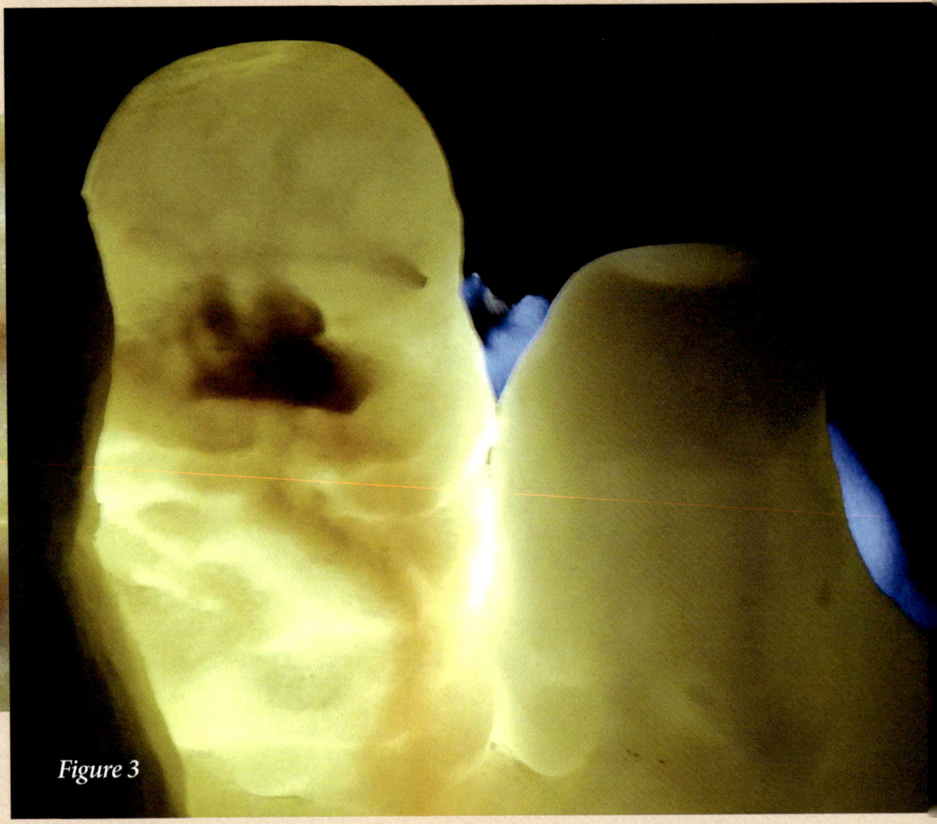

Figure 3

Bats in Rocks

Fossilized animals encased in Carlsbad's speleothems point to a second observation that fits fast speleothem formation. The October 1953 issue of *National Geographic* published a pictorial overview of Carlsbad Caverns. One photo showed a bat's skeleton encased in a stalagmite. This intriguing picture became iconic to those interested in questioning the "thousands of years per centimeter" mantra. The bat fossil demonstrates the speed with which stalagmites can form. How could a bat's intact bones lie undisturbed and undecayed unless its stalagmite formed fast?

Dave Woetzel, co-author of this article, visited Carlsbad in 2021 with the goal of documenting this rock-encased bat. He was surprised to discover more than one such bat (Figure 2). One spot in the Lower Cave at Carlsbad holds special interest (Figure 3). A whole bat lies entombed in a large speleothem that is part of the

Colonel Boles Formation. A guide was able to photograph the bat remains by backlighting the translucent speleothem. And this one suggested more than just bones locked in rock.

Do you see the discoloration in and below the body of the bat? We wonder if decaying flesh and body tissue slid down as the wet stalagmite formed around the bat. Soft tissues decay faster than bone. On the cave floor, cave crickets scavenge the flesh off fallen, dead bats. It looks like this bat got entombed faster than it could get scavenged or decay. While bones might last for thousands of years, softer tissues would have decayed long before the often-cited millennia for speleothem formation would have elapsed.[3]

Fast Stalactites

Bible scoffers have used caves as platforms to propagate old-earth ideas. But actual speleothem formation rates and the bats they encapsulate tell a far different story. We don't need a thousand years to make one centimeter of speleothem. With enough water, such as the Ice Age supplied, even large calcite formations can form in mere centuries. Speleothems can form fast, and bats trapped in rocks show it.

AZTEC RUINS NM
Farmington
CARSON NF
CHACO CULTURE NHP
VALLES CALDERA N PRES
BANDELIER NM
FORT UNION
Santa Fe
PECOS NHP
Las Vegas
Gallup
CIBOLA NF
PETROGLYPH NM
EL MORRO NM
EL MALPAIS NM
Albuquerque
NEW MEXICO
Clovis
Socorro
GILA NF
LINCOLN NF
GILA CLIFF DWELLINGS NM
Roswell
Truth or Consequences
Alamogordo
WHITE SANDS NP
ORGAN MOUNTAINS-DESERT PEAKS NM
Carlsbad
Deming
Las Cruces
CARLSBAD CAVERNS NP
GUADALUPE
LINCOLN NATIONAL FOREST
CARLSBAD CAVERNS NATIONAL PARK
137
180
285

HALEAKALA NATIONAL PARK

ONE OF MANY YOUNG-LOOKING VOLCANOES

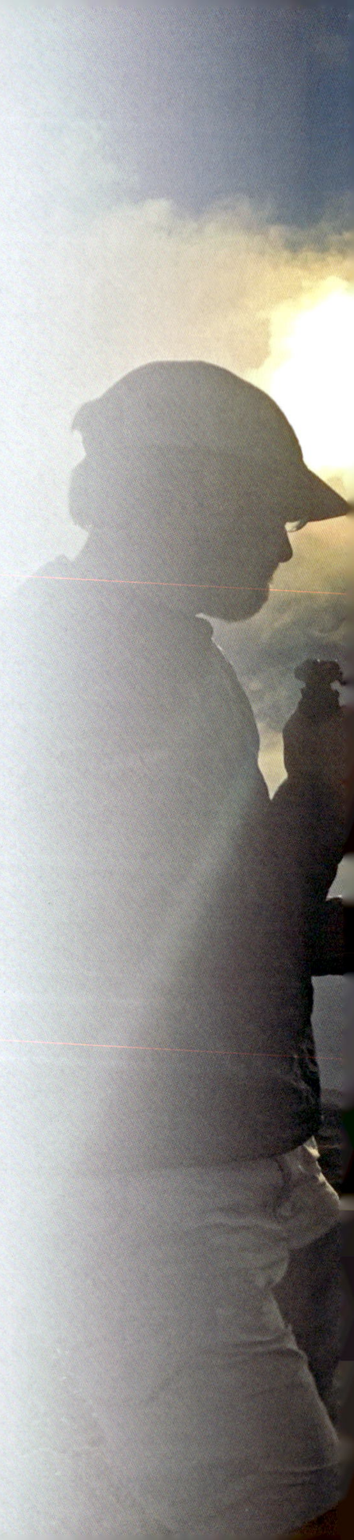

In Hawaii, you can start the day atop the cold peak of Mt. Haleakala and end it on a warm beach on Maui. The sprawling volcanic mountain rises 10,000 feet above sea level. Nēnē birds, cattle, horses, and people populate its slopes and summits. When did Haleakala form?

Maui formed after Oahu and before the Big Island. Hawaiian schoolchildren memorize the standard age assignments. From west to east, Kauai formed supposedly about five million years ago, Oahu three million, Maui Nui one million, and the Big Island 0.4 million. The Big

ICR Video Producer Clint Loveness films the author, Dr. Brian Thomas, near the peak of Haleakala

Figure 1. *The Nēnē* (Branta sandvicensis), *a variety of goose, is the state bird of Hawaii*

Figure 2. *Hawaiian Island chain*

Figure 3. *Dr. Thomas searches for Hawaiian stick spiders (genus* Ariamnes*) on Maui*

Island is still forming today.[1] But three unique observations in and around this national park make Haleakala look only thousands of years old, not millions.

1. Circular Reasoning with Hawaiian Spiders

Spider studies suggest the first clue to a more recently formed Maui. Elusive, nocturnal Hawaiian stick spiders creep among the tropical forests on the mountain's north side. Biologists marvel at how fast these arthropods switch between different body colors. Gold ones live on leaves, white ones on lichen, and dark ones among rocks.

Patterns in their DNA suggest that they pioneered the Hawaiian Islands moving east, starting at Kauai (Figure 2). When a dark stick spider lands on a new island, its offspring suddenly display dark, gold, or white forms. They deploy their color changes faster than "gold and white spiders from the old island have time to reach the new one," according to *Cell Press* via ScienceDaily.[2] These fast flips suggest that internal programming, not many generations of death, sits at the helm of these spider adaptations.

Authors publishing in the journal *Current Biology* framed a spider history using the standard times assigned to the islands as an initial assumption.[3] Does this sound like circular reasoning?

They offered a defense, writing, "Because of the inevitable

Sample	Location	Isotope Age	Actual Age	Reference
Hualalai basalt	Hawaii	1.6 – 0.19 ma*	1800 – 1801 AD	5
Hualalai basalt	Hawaii	22.8 – 16.5 ma	1800 – 1801 AD	6
Kilauea Ika basalt	Hawaii	8.5 – 6.8 ma	1959 AD	6
Kilauea basalt	Hawaii	21 – 8 ma	< 200 years old	7

Table 1

* "ma" is millions of years.

circularity of using island age for calibration, we applied a general arthropod molecular clock estimate of 2.3%/myr."[3] Then they referenced a 1994 study as a basis for their "clock." Problem solved?

Not at all, for the author of that 1994 study wrote, "The estimates of divergence times for these [arthropod] taxa [kinds] are based on dated geological events reported by the authors." Of course, the use of "dated" geological events is just as circular as using "island age" to calibrate timelines, since "island age" equals "dated" geological events.

Circularity is a poor substitute for good science but opens the door to considering a second observation that casts even more doubt on standard dating.

2. Wrong Isotope Ages for Volcanic Rocks

Conventional scientists use radioactive isotopes to estimate ages for certain rocks. The process requires workers to assume how much of which isotope was present when the rock cooled. Some classic tests of this assumption used Hawaiian lava rocks.

The tests were simple: Compare radioisotope age estimates to actual recorded ages. Table 1 shows some results.[4]

Just like these Hawaiian lava rocks, isotope age estimates fail to capture known ages for rocks found around the world. Clear-

ly, we need a whole new way to date rocks. What if we built our world history on what the Lord Jesus—who has always been there—says in His Word? The Bible-based alternative of a flood that reshaped the globe may sound silly to those who first hear it,

***Figure 4.** Dr. Thomas faces Haleakala (top) while investigating lava rocks (bottom)*

Figure 5. *Flat-lying lava layers, uncollapsed lava tubes, and steep cliffsides on Kauai—the oldest of the Hawaiian Islands—signal recent and rapid formation for all the islands, including Maui*

but it fits our next observation: landforms.

3. Three Youthful Landforms

Our hike along Haleakala's Halemau'u Trail revealed flat-lying lava layers. All the Hawaiian islands have these. It looks like the lava landed in sheets, one right after the other. If eons elapsed on Maui between these layers, then where are the ruts and valleys that normal erosion would have etched into the top of each lava layer?

Visitors enjoy Hana Lava Tube northeast of Haleakala. All the islands have lava tubes. The brittle, hollow structures should collapse after only hundreds of thousands of years. The islands and their tubes must have arisen when new seafloors formed near the end of Noah's Flood only thousands of years ago.[8]

Steep-walled cliffs erode faster than other landforms. We peered over steep cliffsides near the top of Haleakala. Indeed, the measured erosion rate of five inches per year would have rounded cliffs and flattened whole islands after a million years.[9] How could these islands keep their cliffs so long?

The next time you visit Haleakala National Park, remember how circular reasoning in spider "ages" and isotope dating failures both cast doubt on millions-of-years thinking. Flat-lying lava layers, fragile lava tubes, and steep cliffs with measured erosion rates point to a Bible-friendly age of thousands of years for the whole world, including its volcanoes.

CRATER LAKE NATIONAL PARK

SERENE BEAUTY AFTER VOLCANIC HISTORY

Stunning Crater Lake, Oregon, with Wizard Island

The deepest and arguably most spectacular lake in the United States is located inside a volcano. Known as Crater Lake, it reaches 1,943 feet at its deepest point.[1] Crater Lake National Park was established in 1902 by President Theodore Roosevelt and became America's fifth national park. Near the center of the Cascade Mountains in south-central Oregon, Crater Lake is about a five-hour drive south from Portland and is eight hours north of San Francisco.[2]

How and when did this lake form? The evolutionary story talks about the Cascade Mountains beginning many millions of years ago.[1] However, this date is based on disproven or unverifiable assumptions and evolutionary dogma. The real account begins with the global Flood just 4,500 years ago.

Flood Origin of Crater Lake

In ICR's Flood model, the Cascade Mountains began development during the receding phase of the Flood.[3] The source magmas were generated as the Pacific seafloor was rapidly subducted or pulled under the West Coast of North America during the Flood year.[4] This process created explosive magmas much different from the Hawaiian volcanoes.[5] This process caused repeated eruptions late in the Flood year and into the Ice Age. The Cascades grew very quickly into massive volcanoes.

The volcano beneath Crater Lake catastrophically erupted for a final time during the Ice Age. Formerly known as Mount Mazama, the 13,000-foot-high volcano blasted out about 75 cubic miles of material.[2] This caused it to collapse upon itself, creating a bowl-shaped crater (caldera). This eruption was 42 times

Merganser duck

more powerful than the 1980 eruption of Mount St. Helens.[2] The highest elevation within the park is now just 8,928 feet at Mount Scott.[1]

Following this final catastrophic explosion, a few lavas flowed out into the open caldera, and finally a tall cinder cone developed. Known as Wizard Island for its cone-shaped appearance, the island rises almost 800 feet above the lake's surface.[1] Cinder cones often form during a final degassing episode (like steam) as volcanoes go dormant or extinct.[6]

Early visitors in 1853 called it Deep Blue Lake, but it was later called Crater Lake in a newspaper account.[1] No streams flow into the lake, and it is only fed by snowmelt from winter snowfalls of about 45 feet each year.[2] Water is only removed by evaporation and groundwater seepage.[1] Because of these factors, the lake is one of the clearest in the world, with visibility down to 120 feet.[1]

Exhibiting Christ's Glory Today

Crater Lake exhibits year-round beauty for those with eyes to see it, and about a half-million visitors view it each year. But some of Christ's creatures don't just visit, they call it home. As a clean freshwater lake, Crater Lake's pure water is habitat to landlocked kokanee salmon and rainbow trout.

Its surrounding shorelands and Wizard Island provide evergreen-forested and felsenmeer (rock-covered) habitat for a variety of large animals—"black bear, bobcat, deer, and marmots"[2]—plus many smaller animals such as pika, chipmunks, and golden-mantled ground squirrels.[7]

Birdwatchers, too, enjoy visiting Crater Lake. Depending upon the time of year, Crater Lake hosts many migratory or resident birds.

> Bald eagles (*Haliaeetus leucocephalos*) and peregrine falcons (*Falco peregrinus*) nest along the caldera cliffs. American dippers (*Cinclus mexicanus*, America's only aquatic songbird) forage at the bottom of fast-flowing streams. Subalpine areas are home to the gray-crowned rosy finch (*Leucosticte tephrocotis*). Wildfire burned forests attract a variety of woodpeckers. . . .Common mergansers (*Mergus merganser*) raise families on the lake, and calls of songbirds permeate the forests and meadows.[7]

Who would expect such a beautiful lake surrounding such a magnificent volcanic cinder cone island? Those who visit Crater Lake and who gaze on Wizard Island can echo the prophet Isaiah's words: "Let them give glory unto the LORD, and declare his praise in the islands" (Isaiah 42:12, KJV).

Yellow-bellied marmot

PICTURED ROCKS NATIONAL LAKESHORE

THE BLANKET SAND FROM NOAH'S FLOOD

Pictured Rocks National Lakeshore lies a few miles east of the town of Munising, Michigan, along one of the most scenic shorelines in the United States. America's first national lakeshore, it was established on October 15, 1966.[1] The park has waterfalls and exposures of massive sand dunes, but it's best known for its nearly vertical multicolored cliffs that rise 50 to 200 feet above Lake Superior. And it turns out these stunning cliffs connect with similar rocks around the world.

The Pictured Rocks' distinctive cliffs consist of colored and stained sandstone. Different minerals give the rocks their various colors. Copper colors them blue or green, iron paints them red or orange, and manganese blackens them. Waters moving underground and discharging create these colors on the exposed cliffs.

Chapel Rock on the Lake Superior coastline

Early Flood Blanket Sands

Nearly 500 miles from the Atlantic Ocean, the Pictured Rocks spectacularly expose some of the earliest Flood sediments, especially sandstones.[2] Called the Munising Formation in Michigan, this same sandstone can be observed across much of the U.S. (Figure 1). It's called the Tapeats Sandstone in Grand Canyon. Because of its great extent and relative thinness, usually less than a few hundred feet, this sand layer is known as a blanket or sheet sand.

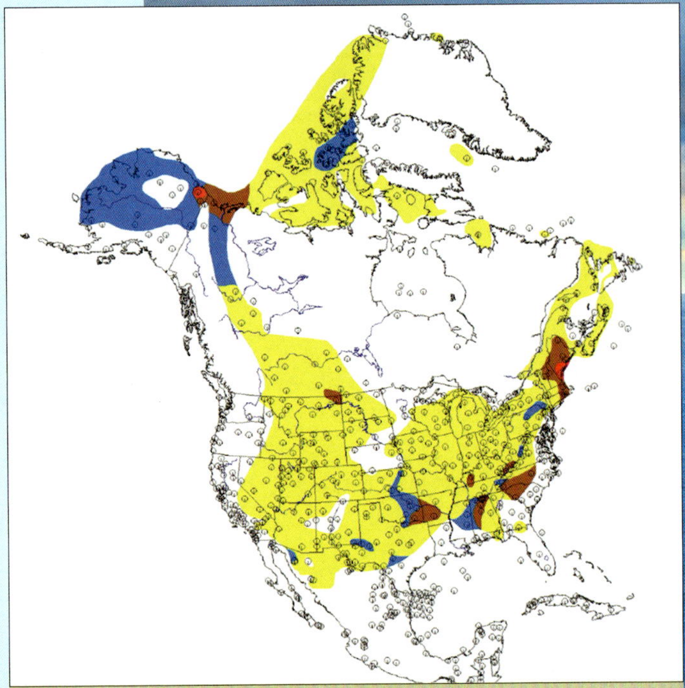

Figure 1. *Basal Sauk map; sandstone is yellow.*

Blanket sands are widespread, thin sandstones spread over large portions of continents. Most are composed of pure, well-rounded quartz grains that contain almost no shale. Uniformitarian geologists have tried to explain their presence for decades and have failed to develop a satisfactory answer.

Some evolutionary scientists have invoked strange conditions that currently exist nowhere to try to account for these blanket sands.[3] Others have claimed that the exact same conditions had to exist at the same time across much of the continents for many millions of years despite the diverse geological environments seen in today's world. Pure, thin sandy layers that blanket the continents should not exist on an old earth.

But we have a better solution. ICR's science staff explains this vast sandstone layer as a consequence of the global Flood. The sandstone is the basal layer of the first megasequence, known as the Sauk Megasequence, that inundated parts of all the continents of the world.[2] We have found that this same sandstone layer exists across the globe. It often coincides with the so-called Cambrian Explosion, where fossils representing most animal phyla suddenly appear in the rock record, further matching the Flood narrative.

Grand Sable Dunes

Munising Falls

Deposition of Sandstones in the Early Flood

The continental-scale blanket sandstones that the Pictured Rocks (Munising Formation) represent were deposited on top of another, less-extensive sandy layer known as the Jacobsville Sandstone. This purple-colored sandstone is visible above the lake level in some parts of Pictured Rocks National Lakeshore. How and when did these layers get here?

Again, the Flood provides the answer. The Jacobsville likely formed in the earliest moments of the Flood as the Midcontinent Rift was bursting open. The Midcontinent Rift represents a tear within the crust of the North American continent running from Michigan in the north to Kansas in the south. Although the rift is only about as wide as the Red Sea, it contains stacked lava flows totaling up to 18 miles thick.[4]

Rift activity probably began when "the fountains of the great deep were broken up."[4] Tremendous volumes of black lava extruded from this rift, which can be seen in the Keweenaw Peninsula of Michigan. The center of the rift created the space for today's Lake Superior. Similar rifting occurred worldwide. The sandstones of the Jacobsville and other sediments were deposited along the rift's edges.

Both the Munising Formation and the Jacobsville Sandstone contain clues that indicate rapid watery deposition during the Flood. These sandstones exhibit stacked layers of cross-bedding, and many surfaces contain ripple marks. These features come from fast-moving shallow waters. The bottom layer of the Munising is known as the basal conglomerate due to its coarse pebbles of rounded quartz and chert.[1] Since it takes more force to move larger grains, pebbles indicate even higher-energy deposition at the base of the Sauk Megasequence.

Bridal Veil Falls

Rapid Erosion and Changing Landscapes

Why are the sandstone cliffs so prominent along the lakeshore? Part of the reason is that the cliffs follow natural vertical fractures in the sandstones that parallel the shoreline.[1] But the presence of sheer cliffs also indicates a youthful origin for the landscape. Again, the recent Flood provides the best solution.

Nearly every year, large chunks of the exposed sandstone break off and fall into Lake Superior.[1] In 2006, one of the "towers" of the iconic Miner's Castle fell into the lake. Wave erosion and freeze-thaw activity expand the natural fractures and cause rock falls, some of which are huge.

The persistence and rapidity of this process is recorded in the numerous rock falls over the historic period, most notably the collapse of Grand Portal Point in 1900. The largest recent event occurred on July 29, 2010, and was captured with a digital camera by professional photographer Lou Waldock from a Pictured Rocks cruise boat.[5]

This rapid erosion indicates these rocks have not been exposed for eons. Mass wasting observed over the last 100 years provides proof. If the Pictured Rocks were exposed for tens of thousands of years or longer, there would be few cliffs left and likely no waterfalls either. If these cliffs are only a few thousand years old, then no wonder we still get to enjoy their beauty.

The sandstones of the Pictured Rocks fit perfectly in the Flood narrative. They were deposited as part of a massive worldwide rise in the ocean level. They exhibit structures like ripples and cross-beds, demonstrating their deposition in high-energy water. And the cliffs look young.

Michigan's Pictured Rocks National Lakeshore showcases the Flood's earliest beginnings in spectacular beauty.

Miner's Castle

The Great Lakes

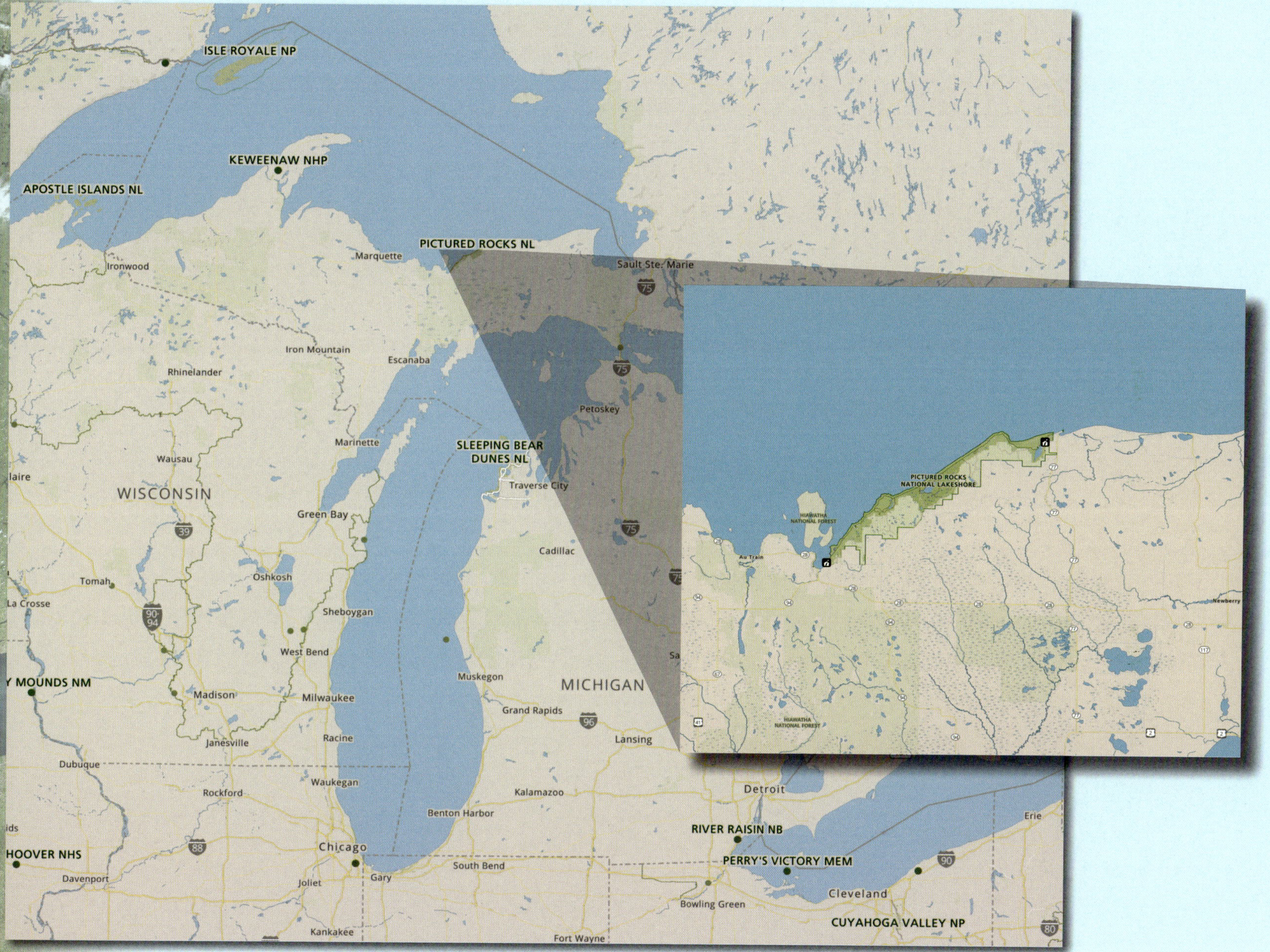
ISLE ROYALE NP
KEWEENAW NHP
APOSTLE ISLANDS NL
PICTURED ROCKS NL
Sault Ste. Marie
Marquette
Ironwood
Iron Mountain
Escanaba
Rhinelander
Petoskey
SLEEPING BEAR DUNES NL
Marinette
Wausau
WISCONSIN
Traverse City
Green Bay
Cadillac
Tomah
Oshkosh
La Crosse
Sheboygan
West Bend
MOUNDS NM
Madison
Muskegon
MICHIGAN
Milwaukee
Grand Rapids
Janesville
Racine
Lansing
Dubuque
Waukegan
Rockford
Kalamazoo
Detroit
Benton Harbor
RIVER RAISIN NB
Erie
HOOVER NHS
Chicago
PERRY'S VICTORY MEM
Davenport
Joliet
Gary
South Bend
Bowling Green
Cleveland
CUYAHOGA VALLEY NP
Kankakee
Fort Wayne
PICTURED ROCKS NATIONAL LAKESHORE
HIAWATHA NATIONAL FOREST
Au Train
Newberry

YELLOWSTONE NATIONAL PARK, PART 1

A FLOOD SUPERVOLCANO

Yellowstone became the world's first national park in 1872. Geologist Ferdinand Hayden led an expedition the year before through much of what became the 2.2-million-acre park, and his report helped convince the U.S. Congress to set aside the land for that purpose. It occupies the northwestern corner of Wyoming and spills into Idaho and Montana.

The park is unique for its travertine flows, mud pots, hot springs, and canyons. It houses over 10,000 thermal features and over half the world's geysers.[1] Its otherworldly landscapes plus raw wildlife have been sufficient causes to protect the region for generations. In these two chapters, we'll explore park features that fit well with biblical history.

Volcanic Beginnings

Yellowstone is one of about 12 supervolcanoes in the world. Thick layers of volcanic ejecta from huge eruptions cover most of the area and beyond. The middle of the park is a volcanic

Grand Canyon of the Yellowstone

The Yellowstone portion of ICR's 2021 science expedition. Left to right: Media Director Michael Hansen, Video Producer Clint Loveness, and Drs. Tim Clarey and Frank Sherwin at the Grand Canyon of the Yellowstone River.

plateau that lies about 7,300 feet above sea level. Yellowstone Lake rests within the plateau, and mountain peaks taller than 11,000 feet surround it.

The central plateau has three overlapping calderas, or collapsed volcanoes. Each collapse occurred after a major eruption. The largest eruption occurred first, and the smallest happened last. This trend reveals diminishing power over time instead of the steady processes that uniformitarians imagine. That scenario fits what we would expect as Earth recovered from the tumultuous Flood about 4,500 years ago.[2]

Geologists noticed a line of older calderas that extends from Yellowstone to Oregon. Conventional scientists consider this trail to have formed from a mantle hot spot as Earth's crust slowly moved across it over millions of years.[3]

However, millions-of-years age assignments rely on an assumed evolutionary timeline, not directly on data. The Institute for Creation Research has a far

Excelsior Geyser Crater

Old Faithful

better interpretation. Evidence continues to build that Earth's tectonic plates moved much faster during the Flood year.[4] Rapid relative movement of North America over a mantle hot spot likely formed the volcanic trail, perhaps within months. Catastrophic conditions slowed dramatically after the Flood. Today, we see little plate motion or volcanism at Yellowstone.

The magma source at Yellowstone is high in silica (quartz-rich). This resulted in especially violent eruptions. In contrast, basalt magmas are lower in silica and flow more gently, like those erupting in Hawaii. Silica-rich continental crust likely altered the deep mantle magmas that rose to the surface at Yellowstone. This unique magma chemistry also helps explain its geysers and hot springs.

Geysers and Hot Springs

Geysers are hot springs that emit jets of hot water and steam. They need an active shallow heat source to superheat groundwater into steam. Yellowstone provides that with magma detected just three miles below the surface.[5]

They also need a fairly watertight plumbing system.[6] The silica-rich magma provides that too. Silica dissolves in circulating hot groundwater. It then drops out of solution (precipitates) onto the fracture surfaces and conduits to form a hard deposit called sinter. This deposit seals water in and makes constrictions that allow geysers to build up pressure.[6]

Water underground expands as it warms until spaces inside the conduits become compressed, like a pressure cooker. Steam and fluids rise, and in places reach Earth's surface at nearly 200°F (93.3°C), sometimes jetting into the sky.[6] Creation scientists speculate how the geysers might illustrate the eruptions through great fractures that took place around the planet when the fountains of the great deep burst open at the start of the Flood (Genesis 7:11).

Old Faithful

Yellowstone's most popular geyser, Old Faithful, may not produce the tallest eruption—although its 130-feet-plus average is impressive—but, as its name implies, it's one of the most predictable. This regularity stems from its steady source of underground water.

But Old Faithful does change. Several decades ago, it averaged less time between eruptions. Local tremors have since lengthened that time.[7] Geysers like Old Faithful point to the role that water played in the formation of Earth and then its reformation during the Flood (2 Peter 3:5-6).

Rainbow Pools

Rainbow-colored hot springs like Grand Prismatic Spring may present the most striking of Yellowstone's natural wonders.

Dr. Brian Thomas (left) and editor Truett Billups at Grand Prismatic Spring on the ICR 2021 science expedition, on which this chapter's images were taken

Different microbial mats form rainbow-like rings when viewed from above. Ferdinand Hayden stated in 1871, "Nothing ever conceived by human art could equal the peculiar vividness and delicacy of color of these remarkable prismatic springs."[8]

Hot water rising from depth cools as it flows from the center of the pools. Heat-loving (thermophilic) microbes called *Synechoccus* use unique biochemistry to pioneer the perimeter of the 370-foot-diameter Grand Prismatic Spring. They live there at about 150°F (65.6°C).[9]

These cyanobacteria harvest sunlight to obtain energy. In the summer, they deploy yellow pigments to protect their vital DNA from the sun's UV rays. Remarkably, they use those same pigments to reassign that would-be-damaging light energy toward their green pigments (chlorophylls) for photosynthesis.

Outside the pool's yellow ring grows a mixture of microbes that together look orange. An even wider red ring lies outside of that. The wider variety of microbes here also use their pigments for protection and light harvesting, but each variation harvests a different part of the light spectrum available for photosynthesis. As the microbes share the light, they convert and store its energy into sugars.

These marvelous microbes form strings, filaments, and mats as they trade resources. Surface dwellers give their sugars to microbes beneath them and receive other nutrients in return. None of this works without precision microengineering, for which an actual engineer like the Lord Jesus deserves full credit.

Grand Prismatic Spring

Mammoth Hot Springs

Mammoth Hot Springs

Hot springs can bring up dissolved lime, sulfur, and even mud. North of Yellowstone's main caldera, Mammoth Hot Springs conveys two tons of dissolved lime from depth per day.[5]

The water here likely gets its heat not from magma like the geysers but from Earth's thermal gradient (increasing temperature with depth) far beneath the surface. The complex at Mammoth Hot Springs covers nearly a square mile and deposits more lime than any other spring in the world.[6]

As the water cools at the surface, lime precipitates to form travertine, a form of limestone, in a complex of terraces. Travertine deposits can grow fast under the right conditions. Extrapolating today's deposition rate into the past shows that this hot spring complex has been active for only thousands of years.

Conclusion

Yellowstone is a beautiful and awe-inspiring reminder of the global Flood. Calderas and thousands of feet of volcanic rock bear witness to great catastrophes that occurred during the Flood year. The magma that powers active geysers has not yet fully cooled, which is what would be expected from the relatively short time since the Flood.

Christ Jesus' creative genius explains the microbes that collaborate to thrive in extremely hot waters. The hot springs display part of the Lord's provision of the hydrologic cycle that moves water around our planet. Although the Flood destroyed the world that then was, it left behind beauty and wonders that point us to Him.

Satellite view of Yellowstone National Park

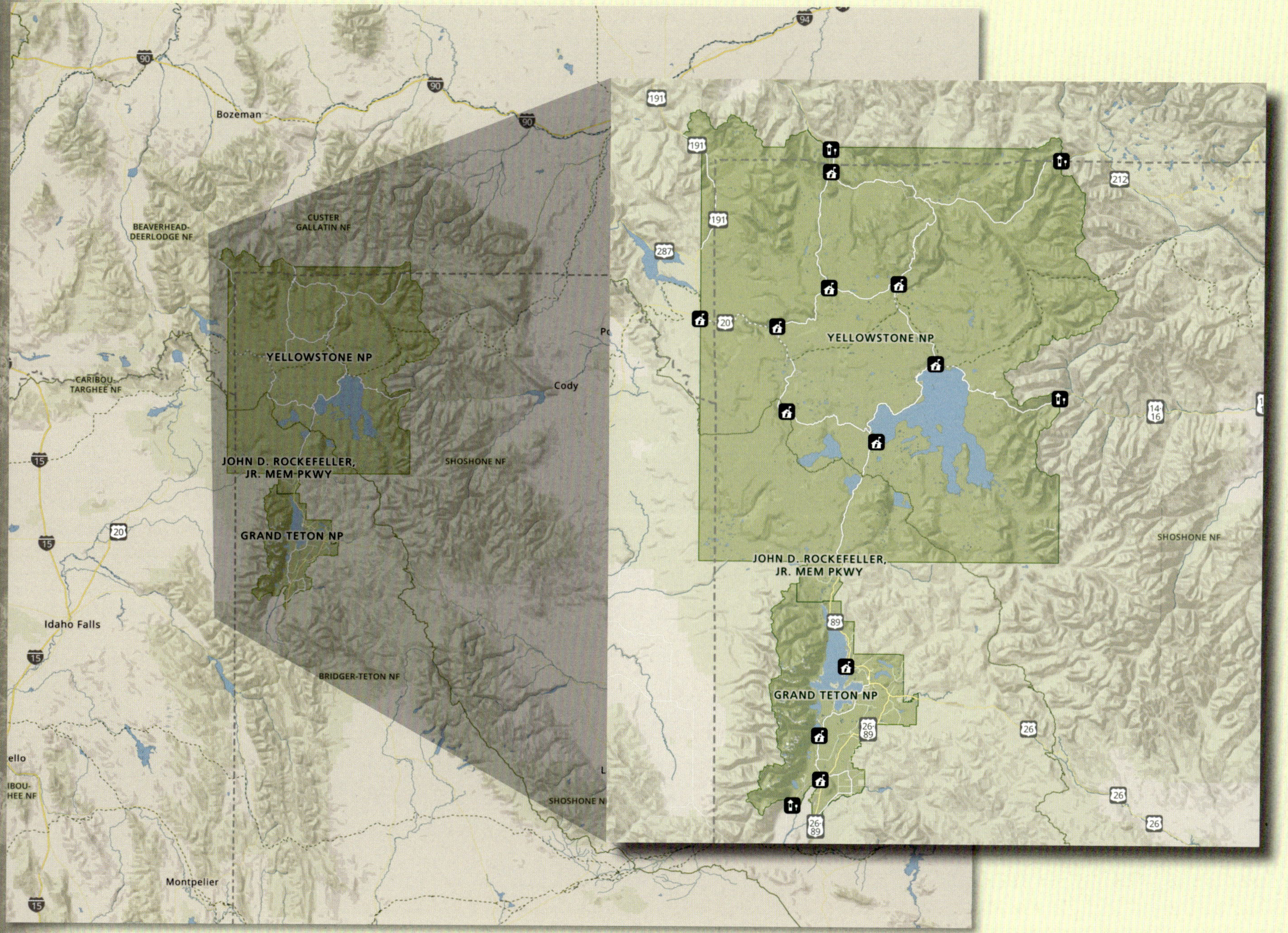
Bozeman
BEAVERHEAD-
DEERLODGE NF
CUSTER
GALLATIN NF
CARIBOU-
TARGHEE NF
YELLOWSTONE NP
Cody
JOHN D. ROCKEFELLER,
JR. MEM PKWY
SHOSHONE NF
GRAND TETON NP
Idaho Falls
BRIDGER-TETON NF
Montpelier
YELLOWSTONE NP
JOHN D. ROCKEFELLER,
JR. MEM PKWY
GRAND TETON NP
SHOSHONE NF

YELLOWSTONE NATIONAL PARK, PART 2

CANYONS AND CATASTROPHE

About three million visitors tour Yellowstone National Park's 3,440 square miles each year.[1] Most come to see natural wonders like Old Faithful and Mammoth Hot Springs, which we featured in the previous chapter.

But this park packs in much more. Yellowstone has its own grand canyon, waterfalls, and even upright petrified trees. Its surface features look young and testify to the catastrophic forces that shaped them.

Grand Canyon of the Yellowstone

A deep canyon with three waterfalls cuts through the eastern spur of the park. This Grand Canyon of the Yellowstone flows for 20 miles and reaches 1,200 feet deep. Before an ancient catastrophe carved the canyon, hot waters moved upward through gray and red volcanic rocks. This hydrothermal action turned them yellow and orange. These yellow stones line the canyon's walls and inspired the park's name and overlook names like Inspiration Point. When did these geologic events occur?

Grand Canyon of the Yellowstone as photographed on the ICR 2021 science expedition

Lower Falls

Water flows from Yellowstone Lake through the Grand Canyon of the Yellowstone over stunning waterfalls. The first cliff forms Upper Falls, dropping a modest 109 feet. Next, the water plummets 308 feet down Lower Falls. Tower Falls' 132-foot drop comes last at the northern end of the canyon.

Conventional geologists believe an Ice Age glacier dammed up a huge lake where Yellowstone Lake is today. When water burst through melted ice, it quickly carved through the rock layers.[2] However, they also believe this river canyon started to form much earlier—about a half million years ago.[3]

Roaring water can erode even hard volcanic rock in a short time. It forms tiny vacuum bubbles that implode with great force in a process called cavitation, which can demolish rock or concrete.[4] After over a half million years, these three waterfalls should have smooth, gentle slopes instead of the steep-sided falls that today command visitors' attention. With Yellowstone's landscape looking only a few thousand years old, it fits well with an Ice Age right after Noah's Flood.[5]

Fossil Tree Trunks and Catastrophe

The northeastern part of the park has petrified tree trunks embedded in volcanic rock debris. A hike up Specimen Ridge reveals some of them. About 20 different layers in the vicinity contain tree fossils, many standing vertically. Sycamore, walnut, chestnut, oak, maple, redwood, and magnolias were fossilized here.[3] Conventional geologists claim these trees grew in place and that over 20 separate eruptions covered as many separate forests over eons.[3] If so, then where are the roots and branches? Instead, some process stripped, sorted, and reburied these tree trunks.

Lessons from the 1980 Mount St. Helens eruption offer a more fitting model. (See the appendix.) That historic blast snapped off a forest of trees at their roots, tore off their branches, and then dumped their trunks into Spirit Lake. Some became waterlogged and sank, thickest end down. They landed upright in the volcanic debris at the lake bottom.[6] An even grander catastrophe must have razed, stripped, and interred the many petrified trees at Yellowstone.

Specimen Ridge

Drs. Tim Clarey and Frank Sherwin discuss Petrified Tree

The Flood's world-destroying violence had the energy to do this. Patterns of rock layers globally and around Specimen Ridge suggest these trees sank soon after the Flood crested on Day 150 (Genesis 7:24). Floodwaters likely ripped them from the high hills. As floodwaters flowed off Earth's surface, they jostled the logs in a moving debris mat. Some tree trunks settled vertically, and some sank faster than others. All the while, Yellowstone's eruptions trapped the trunks in wet volcanic debris and ash. The 20 different layers would have quickly formed one right after the other.

It's a tough hike to Specimen Ridge, but visitors can save their legs by taking a turnout to see Petrified Tree. This big tree grew when Noah was alive around 4,500 years ago. It was torn free by the waves, and the Flood buried it. Later, groundwater deposited minerals inside the wood to petrify it.[7]

Yellowstone's striking canyons and waterfalls showcase a recently carved landscape. Upright petrified tree trunks make stories of multiple forests sound improbable. And the park's clear evidence of the catastrophic forces that shaped it testify to the global Flood recorded in Genesis.

GRAND TETON NATIONAL PARK

THE YOUNGEST ROCKY MOUNTAINS

Just south of Yellowstone National Park lies one of the most picturesque mountain ranges in the Western United States—the Teton Range, established as a national park in 1929.[1] When viewed from the east, the Tetons seem to shoot straight out of the plains. Their stark, jagged peaks rise to heights above 13,000 feet, while the valley floor with its sage and wildlife lies at only 6,000 to 7,000 feet.

Grand Teton National Park's wide variety of animals, including foxes, showcase out-of-this-world innate engineering. Plus, the park's rocks carry clues to a catastrophic start and an icy finish. Let's see where those clues lead.

An image of the Grand Teton National Park mountains taken during ICR's 2021 science expedition

A Fox Superpower

While we were hiking around Jenny Lake at the foot of the Tetons during ICR's 2021 science expedition, a bold fox criss-crossed our trail. During winter, these foxes hunt prey they can't even see. Small mammals crawl beneath several feet of snow, yet foxes find them. How do they know where to look?

Hearing helps, of course. The fox pauses, tilts its head to pinpoint the prey, then takes a bounding leap and plunges, paws and nose first, straight down into the snow. It then emerges with a meal in its mouth. Researchers noticed that a fox finds food an astounding 74% of the time when it attacks "about 20° clockwise of magnetic north."[2]

Foxes facing east or west returned a dismal 18% or less hunting success rate. So, the researchers tested for evidence that foxes have magnetic sensors. The study authors described how the foxes might mentally merge sight, sound, and magnetism.

> A fox moving in a fixed direction when approaching its prey (i.e., moving to the north), could approach until a specific component of the visual pattern generated by the magnetic compass is superimposed on the source of the sound from the prey, so that it could initiate an attack from a fixed distance.[2]

Where could a well-integrated and effective biological system such as this come from? The researchers asserted that it had all somehow evolved. That's like saying it happened by magic. What steps supposedly occurred? What observations even hint that environments, foxes, or any natural condition or combination of natural factors could ever invent something as intricate as a magnetic sensor, let alone combine magnetic, visual, and sound information into a single accurate picture?

Perceptive problem-solvers, not natural processes, craft sensitive instruments like this. These foxes have a superpower. They must have therefore come from a super-powerful Maker.

Making Foxes Friendly...Fast

How many generations do you think it would it take to turn the snarl-and-snap behavior of wild foxes into the friendly approach of domestic dogs? Evolutionary ideas of creature changes demand long periods of time. Charles Darwin wrote:

> For the process of modification and the production of a number of allied forms must be slow and gradual—one species giving rise first to two or three varieties, these being slowly converted into species, which in their turn produce by equally slow steps other species, and so on.[3]

Darwin devotees point to random mutations as the source of changes in DNA that lead to changes in body and behavior. These happen slowly also, building up over many generations.

Evolution trains us to think in terms of long ages.

Researchers tested this question by breeding 130 wild-caught foxes for what they called "tameability." They crossed the friendliest fox parents over many generations. Imagine their surprise when "the aggressive and fear avoidance responses were eliminated from the experimental population in just two-three generations of selection." At the sixth generation, fox pups eagerly sought human contact, complete with wagging tails, "whining, whimpering, and licking in a dog-like manner."[4]

So much for the necessity of ages of mutations. Rapid domestication fits supernatural creation. Why not praise the Lord Jesus since all things, including foxes, "were created through Him and for Him" (Colossians 1:16)?

Catastrophic Beginnings

The rocks that make up the Tetons, the youngest of all the Rocky Mountains, are exposed granite. This rock type probably formed from melt before Noah's Flood, but the sedimentary units that surround it came from the Flood. Catastrophic deposition of about 4,000 feet of early Flood sediments (Paleozoic), 15,000 feet of middle- and high-water Flood rocks (Mesozoic), and up to six miles of receding Flood rocks (Cenozoic) fill nearby areas such as Jackson Hole.[5]

The rock layers that lie deep beneath the ground at Jackson Hole show the same patterns of rocks as similar layers on every continent.[6] Marine-dominated sediments began accumulating as tsunami-driven floodwaters progressed higher and higher from Day 1 to Day 150 of the Flood year. As the waters began to recede after Day 150, they deposited thick Cenozoic layers (Tejas Megasequence).[7] These include volcanic rocks from Yellowstone's eruptions, and thick sands and shales eroded off the rising Rocky Mountains.

The Snake River as it passes east of the Grand Tetons

Stunning Jenny Lake as seen during the ICR 2021 science expedition

The Tetons form a fairly straight, steep mountain front because of a massive, fast-growing fault called the Teton fault.[8] Most of the Rocky Mountains emerged early in the receding Flood as thickened continental crust rose rapidly.[9] But the Teton fault occurred later when the Flood year was nearly over. Like a mammoth trap door, Jackson Hole fell over 15,000 feet while the Tetons rose over 13,000 feet in a matter of days or weeks. These mountains popped up nearly overnight.

The same rocks found on the top of Teton's Mount Moran are found 30,000 feet *beneath* Jackson Hole. What kind of extreme forces accomplished all this geological work? Nothing that fast and violent happens in today's slow processes. Nor do today's processes elevate mountains.

These clues point to unimaginable catastrophe. When did this catastrophe happen? Millions of years should have rounded the Tetons' peaks by now, but they remain tall and sharp. Noah's recent Flood fits all these features.

U-Shaped Valleys and Dammed Lakes

Broad U-shaped valleys separate the individual Teton Range mountains. How and when did these valleys form? Just a few hundred years after the Flood, the Ice Age was at full strength.[10] Ice collected into alpine glaciers on the newly uplifted Tetons. The snake-shaped ice sheets carved out swaths of rock as they slid down the slopes.

Some glaciers deposited piles of rocks, sand, and clay across Jackson Hole, damming the Snake River. Two of these earthen dams formed today's Jackson Lake and Jenny Lake at the foot of the Tetons. The Flood and the Ice Age it caused explain the origin of every main feature of this stunning landscape.

Conclusion

We hold our Bibles in hand when we think about this awe-inspiring park. The Genesis Flood supplied the power to make mountain ranges. Post-Flood conditions led to enough ice to carve the ridges and leave behind the sharp peaks we see today. Finally, creation offers the most sensible answer to the origin of well-integrated and engineered biological systems like acute hearing and magnetic sensors in foxes.[11]

How wonderful to discover such tight matches between science and Scripture in such a beautiful place.

An image of Grand Teton National Park taken on June 15, 2015, by the Operational Land Imager on the Landsat 8 satellite

YELLOWSTONE NATIONAL PARK
To West Thumb
South Entrance
Flagg Ranch Information Station
Headwaters Lodge & Cabins at Flagg Ranch
Ashton - Flagg Ranch Road
CARIBOU-TARGHEE NATIONAL FOREST
JOHN D. ROCKEFELLER, JR. MEMORIAL PARKWAY
Glade Creek Trailhead
GRAND TETON NATIONAL PARK
BRIDGER-TETON NATIONAL FOREST
Lizard Creek
Leeks Marina
Colter Bay Village
Colter Bay Visitor Center
Hermitage Point Trailhead
JACKSON LAKE
Two Ocean Lake Trailhead
Jackson Lake Lodge
Jackson Lake Junction
Signal Mountain Lodge
Moran Junction
Pacific Creek
Buffalo Valley Road
To Dubois
Hatchet
TETON RANGE
Teton Park Road
North Jenny Lake Junction
Jenny Lake Lodge
JENNY LAKE
South Jenny Lake Junction
Jenny Lake Visitor Center
YOU ARE HERE
Cunningham Cabin Historic Site
Triangle X Ranch
Grand Targhee Resort
Teton Canyon
CARIBOU-TARGHEE NATIONAL FOREST
Driggs
TETON
JACKSON HOLE
ANTELOPE FLATS
Antelope Flats Rd
Mormon Row Historic District
Craig Thomas Discovery and Visitor Center
Moose Junction
Gros Ventre Road
Atherton Creek
Crystal Creek
Gros Ventre
BRIDGER-TETON NATIONAL FOREST
Laurance S. Rockefeller Preserve
Granite Canyon Trailhead
Teton Village
Jackson Hole Mountain Resort
Gros Ventre Junction
NATIONAL ELK REFUGE
GROS VENTRE RANGE
Victor
To Idaho Falls
Mike Harris
Trail Creek
BRIDGER-TETON NATIONAL FOREST
Teton Pass
Wilson
Jackson National Fish Hatchery
National Museum of Wildlife Art
Jackson Hole and Greater Yellowstone Visitor Center
JACKSON
Snow King Resort
To Pinedale
Curtis Canyon
IDAHO
WYOMING
North
5 Kilometers
5 Miles
Paved road
Unpaved road
Trail
Ranger station
Self-guiding trail
Horseback riding
Marina
Food service
Picnic area
Store
Gas station
Lodging
Campground
Tent-only campground

View of Mackinac Island's harbor from Fort Mackinac

MACKINAC ISLAND STATE PARK

FLOOD, ICE AGE, AND...DIGESTION?

Originally designated in 1875 as America's second national park, Mackinac Island's federal land was transferred to the State of Michigan in 1895 and became its first state park.[1] It remains one of the top tourist destinations in the state. Pronounced "mak-in-aw," the 3.8-square-mile island is nestled between Michigan's Upper and Lower Peninsulas on the Lake Huron side of the nearly five-mile-long Mackinac Bridge. Known for its lack of automobiles, the island boasts a Revolutionary War-era fort and the Grand Hotel, with a 660-foot porch.

But the story of the island goes back to the global Flood, with rocks exposing sediments from the earliest weeks of that cataclysm. Post-Flood wave-cut cliffs preserve a history of fluctuating lake levels during the Ice Age. And an accident on the island involving a bullet wound that didn't

heal revealed God's design in the human digestive system.

Flood Beginnings and Collapses

Mackinac Island exposes rocks from two of the earliest megasequences of the Flood, the Tippecanoe and the Kaskaskia. The Kaskaskia is the younger of the two and lies on top of the Tippecanoe. Globally, these layers are filled with shallow-water marine fossils since at that point the floodwaters had not yet begun to cover the dry land.[2]

Some of these rocks are visible near the center of the island on the highest point. Referred to as the Ancient Island, its peak is about 275 feet above today's Lake Huron.[1] Here, near the War of 1812-era Fort Holmes, some of the lower Kaskaskia sediments are exposed, known as the Devonian Bois Blanc Formation. Below this half-mile-long hill and across the bulk of the island reside carbonate rocks of the upper Tippecanoe Megasequence (Silurian St. Ignace Dolomite). These two sedimentary layers make up the bulk of the island and were most likely deposited within the first 40 days of the Flood.[2] They were exposed as

Sugar Loaf, one of the largest spires of Mackinac Breccia, with Dr. Tim Clarey and one of his grandsons at the base

the floodwaters receded and were further sculpted by the post-Flood Ice Age.

Scattered around the island are several prominent "sea" stacks composed of broken and angular pieces of rock. These resistant spires, stacks, arches, and even a cave are composed of Mackinac Breccia. Breccia is made up of broken, angular chunks of rock cemented in sediment. The largest stack of Mackinac Breccia is called Sugar Loaf and is about 75 feet tall. Arch Rock is a rare formation of limestone breccia and stands on the Lake Huron shoreline 146 feet above the water.

These stacks formed in a manner similar to a sinkhole and were the product of salt dissolution and collapse. In this case, the sediments composing the broken pieces were deposited on top of the St. Ignace Dolomite. Below this dolomite lie earlier Flood rocks that are rich in salt.

During the Flood's continued advance, water was forced into the ground and dissolved some of the salt. With no support from below, the overlying sediments collapsed, dropping fragments from above into the vacant cavity. Swirling groundwater cemented these bro-

Arch Rock

Old Fort Mackinac showing the wave-cut cliff representing the Ice Age Nipissing Lake level. Note the breccia spire on the right.

ken fragments together, and, rather oddly, they became more resistant than the surrounding bedrock.

As the receding Flood stripped away the surface layers, the breccia stacks were left behind as "sea" spires. Glacial lake waters further carved some of these into caves such as Skull Cave and arches such as the previously mentioned Arch Rock.

Rapidly Carved Ice Age Cliffs

Native Americans referred to the island as Michilimackinac, or "place of the great turtle," because in profile the island resembles a turtle's back.[3] This appearance is due to erosion during the two primary Ice Age lake levels, the older Lake Algonquin and the younger Lake Nipissing. The Ancient Island was all that was exposed above the water line when Lake Algonquin crested at over 230 feet above the present lake level. Its waves carved a steep cliff marking the edge.

As the lake levels further dropped, over two dozen successive shoreline ridges were formed prior to the formation of Lake Nipissing. This younger level was only about 50 feet above the level of Lake Huron today. It also formed a 45- to 50-foot wave-cut cliff that nearly circles the island. Fort Mackinac was established on this cliff in 1780, as was the Grand Hotel in 1887.[3]

The two prominent wave-cut cliffs and the numerous shoreline ridges preserve evidence of the many changing Ice Age lake levels. These features always seem to be present wherever Ice Age lakes resided. In fact, similar wave-cut terraces are found surrounding the Great Salt Lake in Utah (ancient Lake Bonneville) and around the glacial Lake Missoula in western Montana.

Wave-cut cliffs don't take thousands or even hundreds of years to develop. In fact, these landforms can occur in just a few weeks.[4] Creation scientists calculate that the Ice Age lasted until about 500 to 700 years after the Flood.[5] The development of these landforms easily fits within the biblical timeline.

Design in Human Digestion Revealed

Mackinac Island is also famous for a digestive-related reason. And it's not for the island's delicious fudge that's made each summer to fuel hungry tourists. Rather, it's because an accident on the island led to some of the earliest studies of human digestion.

It began with a blast in 1822 in the American Fur Company store on the is-

William Beaumont

Skull Cave

land's Main Street. A gun accidentally discharged into the abdomen of Alexis St. Martin, a young French Canadian. Fort Mackinac's post surgeon, Dr. William Beaumont, was called. After St. Martin was patched up, he was left with a hole, or fistula, in his stomach that didn't fully heal. In 1825, Dr. Beaumont decided to do some experiments on his young patient since at the time no one knew much about the digestive process.

> Beaumont commenced his experiments in May at Fort Mackinac. As St. Martin lay on his right side, the doctor peered into his stomach and observed the stomach's reaction to food and drink. He siphoned out fluids and retrieved food with a spoon to see how it had changed since entry. Frequently he suspended pieces of raw or roasted meat on string to find out how long digestion took. Once he substituted a plug of beef for cloth over the orifice; five hours later, he noted it had been "completely digested off, as smooth and even as if it had been cut with a knife."[6]

Dr. Beaumont conducted a total of 250 experiments on St. Martin that first year. He eventually published the results of his groundbreaking discoveries in a book on human digestion.[7] A horrible accident on a remote island and an inquisitive army doctor who had a lifelong faith in God together unlocked some of the mysteries of our wonderfully designed human body.

Mackinac Island State Park thus showcases the Flood, the Ice Age, and the engineering prowess of our Creator, the Lord Jesus Christ.

NATIONAL FOREST
East Lake
Ozark Creek
Pine River
Munuscong River
East Branch Munuscong River
Brevoort Lake
MARQUETTE ISLAND
HIAWATHA NATIONAL FOREST
Lake Huron
MACKINAC ISLAND
HIAWATHA NATIONAL FOREST
BOIS BLANC ISLAND
Lake Paradise
Cheboygan
Black River
Douglas Lake
HIAWATHA NATIONAL FOREST
North Country National Scenic Trail
Mullett Lake
Burt Lake
Black Lake
Crooked Lake
Petoskey
Pigeon River
Sturgeon River
Black River
Rainy River
Walloon Lake
Lake Charlevoix

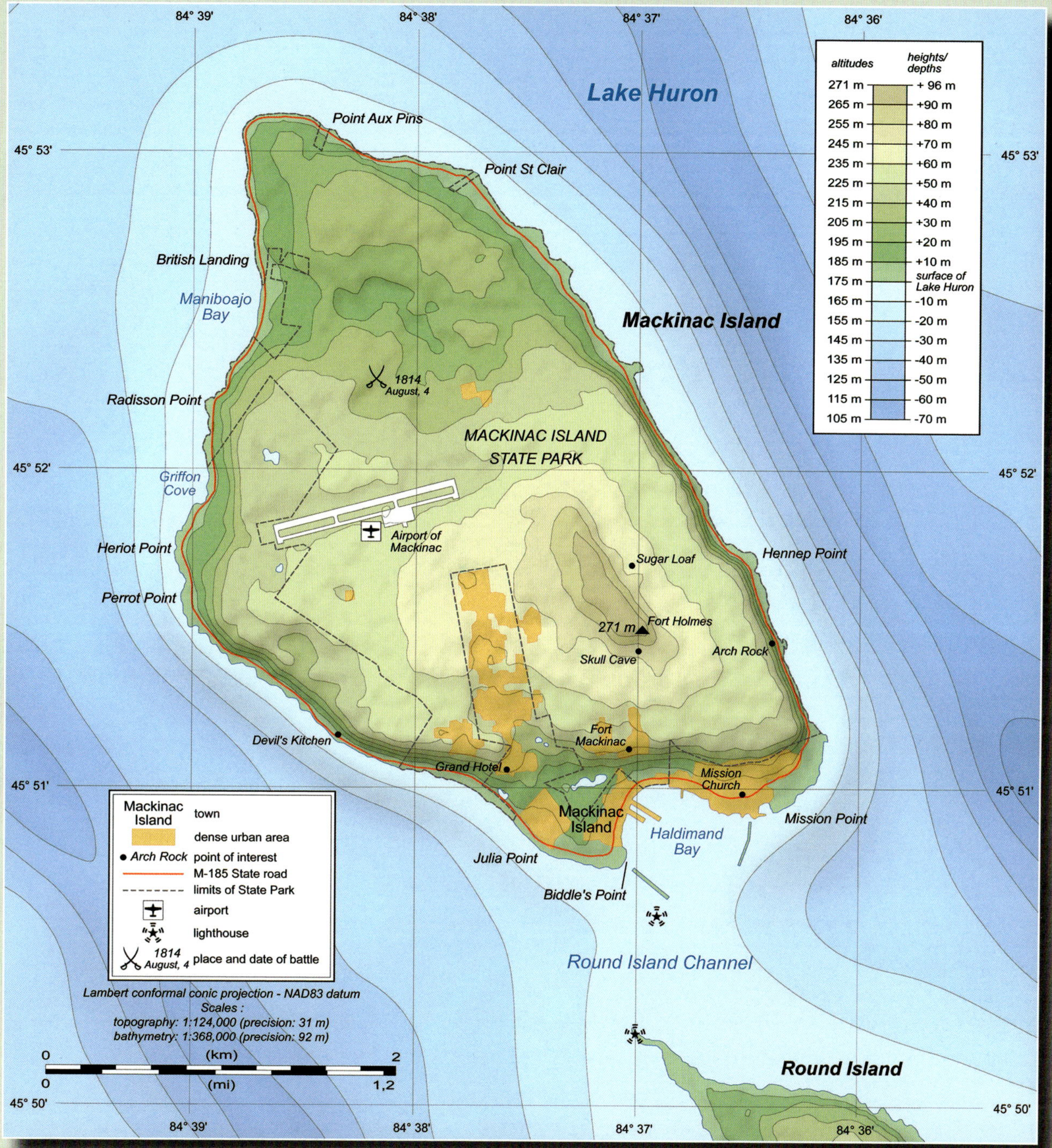
Lake Huron
Mackinac Island
MACKINAC ISLAND STATE PARK
Point Aux Pins
Point St Clair
British Landing
Maniboajo Bay
Radisson Point
1814 August, 4
Griffon Cove
Heriot Point
Airport of Mackinac
Perrot Point
Sugar Loaf
Hennep Point
271 m
Fort Holmes
Skull Cave
Arch Rock
Devil's Kitchen
Fort Mackinac
Grand Hotel
Mission Church
Mackinac Island
Mission Point
Haldimand Bay
Julia Point
Biddle's Point
Round Island Channel
Round Island
altitudes
heights/ depths
271 m + 96 m
265 m +90 m
255 m +80 m
245 m +70 m
235 m +60 m
225 m +50 m
215 m +40 m
205 m +30 m
195 m +20 m
185 m +10 m
175 m surface of Lake Huron
165 m -10 m
155 m -20 m
145 m -30 m
135 m -40 m
125 m -50 m
115 m -60 m
105 m -70 m
Mackinac Island town
dense urban area
Arch Rock point of interest
M-185 State road
limits of State Park
airport
lighthouse
1814 August, 4 place and date of battle
Lambert conformal conic projection - NAD83 datum
Scales :
topography: 1:124,000 (precision: 31 m)
bathymetry: 1:368,000 (precision: 92 m)
0 (km) 2
0 (mi) 1,2
84° 39'
84° 38'
84° 37'
84° 36'
45° 53'
45° 52'
45° 51'
45° 50'

APPENDIX

MOUNT ST. HELENS A LIVING LABORATORY

Nothing put a damper on uniformitarianism like the Mount St. Helens eruption on May 18, 1980. That explosion still echoes through the halls of the scientific establishment over 40 years later. For nearly 150 years prior to the eruption, strict uniformitarianism reigned supreme in geology. The influence of James Hutton and his concept of deep time had trickled down to even the smallest details. Every geological process was thought to proceed as slowly as those observed today. Erosion and deposition were seen as steady, methodical processes requiring vast amounts of time to make a substantial impact.

In 1980, Mount St. Helens dropped an outdoor laboratory in geologists' laps, forcing them to accept catastrophic events as major contributors to Earth's overall geologic story. Many geolo-

View into the Mount St. Helens crater from Johnston Ridge. Note the Little Grand Canyon in the foreground.

Eruption of 1980

gists call this actualism as opposed to uniformitarianism.[1] They now accept the evidence that catastrophic events make major impacts on the rock record and that the normal everyday processes of deposition and erosion contribute very little.

Mount St. Helens has even impacted the science of biology. Recovery by flora and fauna in the devastated area around the volcano occurred very quickly. Many centuries aren't required to turn a lifeless terrain into a lush biota filled with life. Plant and animal repopulation are remarkably fast. All of this has implications for the worldwide Flood recorded in Genesis.

Rapid Deposition

The cause of the sudden shift in thought was scientists' direct observation of the effects of Mount St. Helens' volcanic activity. Geologists documented that up to 400 feet of new strata have formed at the volcano since the first eruption in 1980.[2] These deposits originated from air fall, pyroclastic flows, landslides, and even stream water. The geologists saw that laminated deposits (thin layers) can be produced quickly. Previously, laminated strata were believed to take many years to form, with possibly one layer laid down each year. We now know this assumption is false. One deposit at Mount St. Helens resulted in the creation of a 25-foot-thick finely laminated unit in a matter of hours![2]

Little Grand Canyon was carved in one day

New studies show that rapid deposition is the norm, not the exception. Conventional science has used the slow deposition of sediments like clay and lime mud (micrite) as an argument for an old earth, claiming that all clays form by slowly settling out of stagnant water. People have been indoctrinated with the notion that enormous periods of time are necessary to explain these thick rock layers.

We do see clay settling out of stagnant water today, but the clay-rich rocks we observe didn't form that way. Rocks like shale and mudstone often exhibit fine laminations a few millimeters thick. These layers didn't result from deposition in stagnant water. Empirical evidence demonstrates that laminated clays must be deposited in energetic settings by moving water.[3] The results match the predictions of creation geologists who interpret clay, which forms mudstones and shales, as rapid deposits that occurred during the year-long Flood.[4]

Dr. Tim Clarey at Spirit Lake

A second finding also has uniformitarian geologists bewildered. Although some lime-rich rocks called carbonates have been interpreted as forming in high-energy settings, carbonate mud has always been thought of as forming in "quiescent ocean settings."[5] But laboratory studies show that micrite is also deposited by moving water. Laminated limestones, like laminated mudstones, aren't the result of a slow settling process, as was previously thought.

According to the authors of a 2013 study, "these experiments demonstrate unequivocally that carbonate muds can also accumulate in energetic settings." They added, "Observations from modern carbonate environments and from the rock record suggest that deposition of carbonate muds by currents could have been common throughout geologic history."[6]

Rapid Erosion

Mount St. Helens also demonstrated that erosion can be much quicker than previously claimed. The eruption's steam blast, ash flows, and volcanic mudflows rapidly changed the landscape surrounding the volcano and its waterways. The North Fork of the Toutle River had to carve a new course because the 1980 eruption blocked the original passage with nearly a cubic mile of debris.[2]

After a small subsequent eruption on March 19, 1982, a mudflow from melted snow and ice flowed down the North Fork of the Toutle River Valley, carving a new canyon up to 140 feet deep.[2] This "Little Grand Canyon" is an approximately 1/40th-scale version of Grand Canyon, demonstrating the rapid scouring power of water. Creation geologists frequently use this analogy to explain the rapid formation of the much larger Grand Canyon. Erosion can be fast under the right conditions, and creation geologists believe the global Flood provided ample water to carve canyons and erode mountains in a short amount of time. This was especially true in the receding phase of the Flood as water energetically washed off and through the soft-sediment-laden landmasses into the newly forming ocean basins during continental uplift.

In fact, it's been well known for decades that even uniformitarian rates of erosion are still so fast that the continents themselves should have been reduced to sea level long ago.[7] A 2011 study confirmed that the average erosion rates of exposed rocks are about 40 feet per million years.[8] This would completely erode

Spirit Lake, 1982 (top). "Whaleback" feature in the crater, 2005 (bottom).

most continents in less than 50 million years, and yet they still exist.

Rapid Removal of Vegetation

The 1980 eruption of Mount St. Helens sent the largest landslide ever recorded down the north side of the mountain. Over 3.3 billion cubic yards of rock and ice, moving at speeds exceeding 150 mph, tore the side of the mountain open, unleashing a devastating steam blast. About 25% of this material ended up in Spirit Lake.[9]

The 680 million cubic yards of material that hit Spirit Lake formed a gigantic tsunami that ripped across the hillsides north of the lake.[9] This water wave sheared off an estimated one million fully grown trees near their bases and transported the logs back to the lake as the water receded. Many of these trees have been found floating upright with the root end down. Following a survey of the floating trees in 1985, it was estimated that more than 19,000 upright logs had settled on the floor of the lake.[2]

Dr. Harold Coffin of the Geoscience Research Institute found that many of the upright trees were randomly spaced, not clumped together, across the bottom of the lake.[2] And he noted that many of the trees had settled at various levels in the mud, giving the appearance of deposition at different times.

Creation geologists use Spirit Lake's floating log mat and sunken logs as an analogy for the likely devastation that occurred during the Flood year. The sunken upright trees are used to explain the numerous polystrate trees often found extending through coal beds. They also help explain petrified forests like the one at Specimen Ridge in Yellowstone National Park. There we find upright trees at different stratigraphic levels that could have formed all at the same time during the global Flood.

Today, over 40 years later, a massive log mat remains floating around Spirit Lake. However, if this mat had been rapidly buried by subsequent sediments, it's likely these trees would have turned into a coal bed. Coal beds also don't take vast amounts of time to form; they just need the right conditions.

Rapid Recovery of Flora and Fauna

When Noah and his family looked out on the bleak and barren post-Flood landscape, they must have wondered how long it would remain that way. After a natural disaster, creation scientists observe environmental recovery processes and extrapolate to the worldwide reclamation after the Flood. Mount St. Helens continues to be a scale model of the world God destroyed and reformed as a result of His judgment.

Scripture states that "all the fountains of the great deep were broken up" (Genesis 7:11) as the Flood began. This breakup most likely included worldwide volcanic activity that

Logs still float on Spirit Lake over 40 years after the 1980 eruption

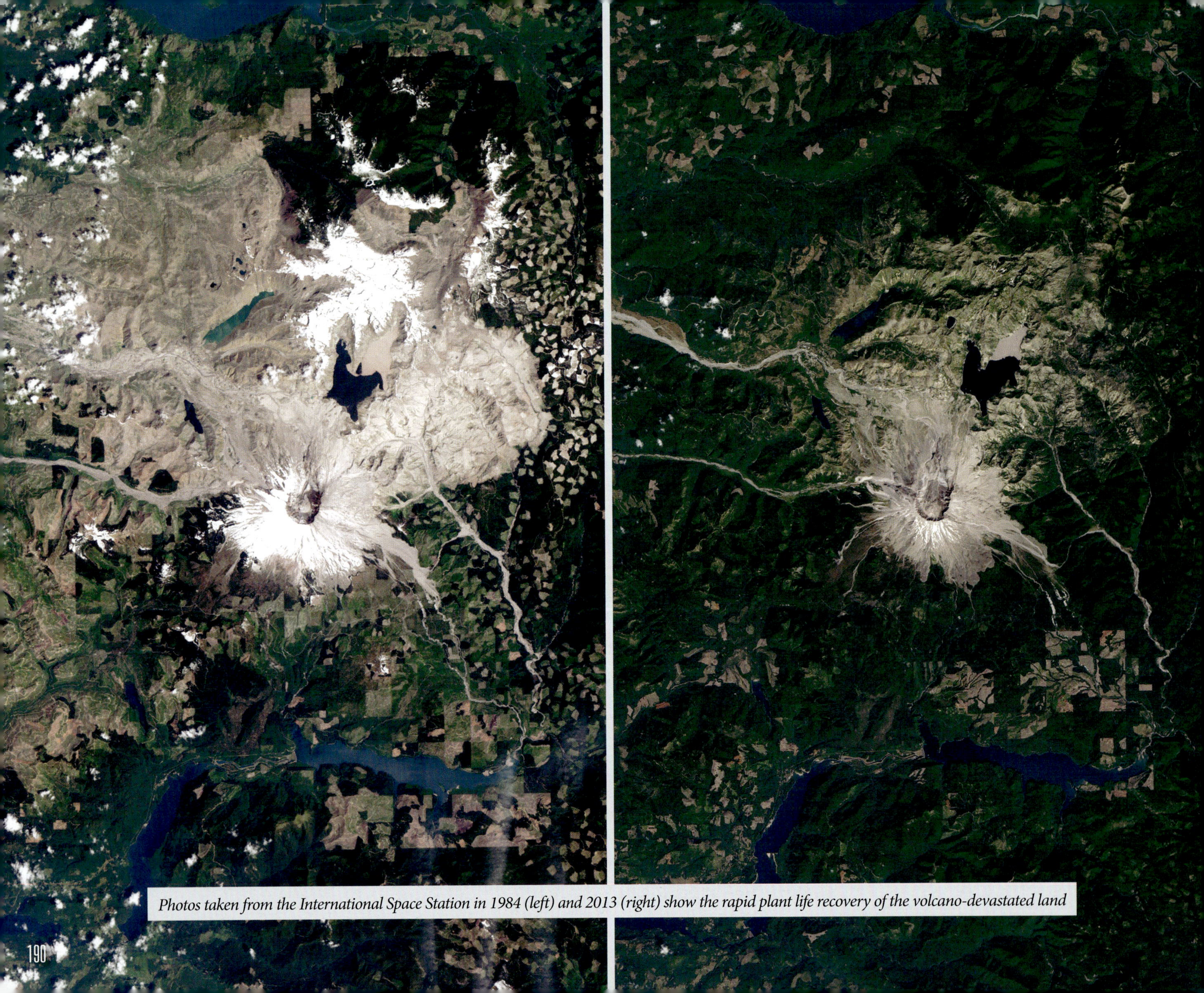

Photos taken from the International Space Station in 1984 (left) and 2013 (right) show the rapid plant life recovery of the volcano-devastated land

continued all over the earth for 150 days (Genesis 7:24; 8:2). Imagine the devastation!

Could there be any recovery after such wholesale destruction? Surtsey is a good example of a mature and varied landscape that developed in just a few months following the island's volcanic formation in the Atlantic Ocean in 1963.[10] A life scientist who studied it stated in 2008, "Surtsey always provides surprises....We discover about 20 new species [of life forms] each year."[11] Together, about 60 plant species including mosses, lichens, and an evergreen shrub have been established since Surtsey formed. Its rapidly growing ecosystem is powerful evidence against critics who claim that Earth could not possibly have recovered yet from a worldwide flood that happened only thousands of years ago.

In 2015, research was published[12] regarding "river ecosystems [that] show 'incredible' initial recovery after dam removal" in the western United States.[13] A related article stated:

> During his time conducting the studies in Washington, [ecologist Christopher] Tonra watched reservoir beds that looked like moonscapes return to vibrant, rich habitat and cascades emerge where none had been, at least for the last century. "Watching that happen was just incredible," he said.[13]

The most significant volcanic upheaval in 20th-century America witnessed a similar remarkable renewal of a devastated environment.

> When Mount St. Helens erupted in 1980, it destroyed every living thing around it. Gas, ash and rock, heated to over 1000 degrees Fahrenheit, sterilized a 60-kilometer square area, leaving a gray lunar-looking landscape devoid of plants

and animals. Within a year, the first plant life had started to return, just as ecologists predicted it would.[14]

In fact, "the recovery of the Mount St. Helens area was 'a wonderful living laboratory' to investigate how ecosystems and species respond to and recover from major disturbances, said Charlie Crisafulli, a research ecologist."[15] Biologists at the University of Washington observed the recovery of two areas covered by violent types of debris flows or mudflows called lahars. They found "there are striking differences—the forest-surrounded lahar has recovered much faster and has pines and firs atop it, while the more isolated lahar is still mostly covered by grasses, early-stage colonizers."[15]

The Mount St. Helens area has quickly recovered from intense devastation. After only 20 years, biologists noted a rapid recovery of animals and plants on what had been something close to a thermonuclear blast zone. Today, the 40-plus-year-old zone is a lushly treed forest. Noah and his family no doubt

Recent view of the lava dome and the ice buildup in the crater of Mount St. Helens

witnessed the same kind of rapid recovery in the decades following the global Flood.

Conclusion

Mount St. Helens has provided over 40 years of empirical data that support catastrophism and refute strict uniformitarianism. The eruptions have even changed the way conventional scientists view Earth's processes, shifting them to be more accepting of catastrophism. Creation scientists will continue to use Mount St. Helens as a living laboratory to study the devastating effects of events like the global Flood and Earth's rapid recovery in miniature. It is truly a lasting monument to catastrophe.

REFERENCES

Grand Canyon: Exposing the Flood

1. An unconformity is where some amount of erosion has taken place, removing a segment of the rock record. Unconformities can also occur if there was a brief hiatus in deposition. Unconformities do not imply vast amounts of time since they can form quickly as tsunami-like waves wash across the landscape, eroding the surface as they go.
2. Morris, J. and S. A. Austin. 2003. *Footprints in the Ash: The Explosive Story of Mount St. Helens*. Green Forest, AR: Master Books, 72-76.
3. Austin, S. A. 1986. Mt. St. Helens and Catastrophism. *Acts & Facts*. 15 (7).
4. Clarey, T. 2018. Grand Canyon Carved by Flood Runoff. *Acts & Facts*. 47 (12): 10-13.
5. Clarey, T. 2020. Lava Flows Disqualify Lake Spillover Canyon Theory. *Acts & Facts*. 49 (10): 10-12.

Arches National Park: Sculptures from the Flood

1. Johnson, D. W. 2016. *Arches: The Story Behind the Scenery*. Las Vegas, NV: KC Publications, 14.
2. Portenga, E. W. and R. R. Bierman. 2011. Understanding Earth's eroding surface with ^{10}Be. *GSA Today*. 21 (8): 4-10.
3. Bruthans, J. et al. 2014. Sandstone landforms shaped by negative feedback between stress and erosion. *Nature Geoscience*. 7: 597-601.
4. Lovett, R. A. Sandstone arches form under their own stress. *Nature News*. Posted on nature.com July 20, 2014, accessed November 19, 2020.

Gunnison's Black Canyon: The Flood Solves Mysterious MIssing Time

1. Not to be confused with literal fossilized lightning strikes called fulgurites, branching formations of quickly melted and re-solidified materials.
2. Zaenger, P. 2016. *Black Canyon of the Gunnison: The Story behind the Scenery*. Wickenburg, AZ: KC Publications, 11.
3. Portenga, E. W. and R. R. Bierman. 2011. Understanding Earth's eroding surface with ^{10}Be. *GSA Today*. 21 (8): 4-10.
4. This question assumes against evidence to the contrary and even against the principle of uniformity that past erosion rates were much slower than today's in order for continental rock to still exist above sea level. See Thomas, B. Continents Should Have Eroded Long Ago. *Creation Science Update*. Posted on ICR.org August 22, 2011.
5. Conventional interpretations assume the giant sand dunes within the Entrada formed from wind. However, cross-beds form close to an angle of 20° under flowing water and 32° under moving air, and the Entrada dunes measure closer to 20°.
6. Clarey, T. 2015. Dinosaur Fossils in Late-Flood Rocks. *Acts & Facts*. 44 (2): 16.
7. Clarey, T. 2020. *Carved in Stone: Geological Evidence of the Worldwide Flood*. Dallas, TX: Institute for Creation Research, 283-311.

The Painted Desert: Fossils in Flooded Mud Flats

1. Sadler, K., W. Parker, and S. Ash. 2015. *Dawn of the Dinosaurs*. Petrified Forest, AZ: Petrified Forest Museum Association.
2. Ibid, 35.
3. Ibid, 45.
4. Plummer, C. C., D. McGeary, and D. H. Carlson. 2005. *Physical Geology*, 10th ed. New York: McGraw Hill, 230.
5. Sadler et al, *Dawn of the Dinosaurs*, 86.
6. Ibid, 92.
7. Ibid, 48.
8. Chatterjee, S. 2015. *The Rise of Birds: 225 Million Years of Evolution*, 2nd ed. Baltimore, MD: Johns Hopkins University Press.
9. Tomkins, J. P. 2021. The Fossils Still Say No: The Fins-To-Feet Transition. *Acts & Facts*. 50 (3): 10-13.
10. Clarey, T. 2020. *Carved in Stone: Geological Evidence of the Worldwide Flood*. Dallas, TX: Institute for Creation Research, 256-281.
11. Clarey, T. 2015. Dinosaurs in Marine Sediments: A Worldwide Phenomenon. *Acts & Facts*. 44 (6): 16.
12. Clarey, T. 2019. Marine Fossils Mixed with Hell Creek Dinosaurs. *Acts & Facts*. 48 (4): 10.

Zion National Park: Evidence of Deep Water Sand Waves

1. Zion National Park: Sandstone Cross-bedding. National Park Service. Posted on nps.gov, accessed March 8, 2021.
2. Weaver, L. Navajo Sandstone. Posted on UtahGeology.com April 16, 2015, accessed March 10, 2021.
3. Allen, J. R. L. 1970. *Physical Processes of Sedimentation*. New York: American Elsevier, 78. Cited in Austin, S. A. 1994. *Grand Canyon: Monument to Catastrophe*. San Diego, CA: Institute for Creation Research, 33.
4. Parrish, J. T. and H. J. Falcon-Lang. 2007. Coniferous tree associated with interdune deposits in the Jurassic Navajo Sandstone Formation, Utah, USA. *Palaeontology*. 50 (4): 829-843.
5. Clarey, T. 2015. Dinosaur Fossils in Late-Flood Rocks. *Acts & Facts*. 44 (2): 16.
6. Clarey, T. 2020. *Carved in Stone: Geological Evidence of the Worldwide Flood*. Dallas, TX: Institute for Creation Research, 114-151.

The Everglades: Two Biology Basics the Bible Got Right

1. Parker, G. 2011. *Building Blocks in Life Science*. Green Forest, AR: Master Books, 59.
2. Virata, J. Burmese Pythons Crossbreeding With Indian Pythons In Florida Everglades. *Reptiles*. Posted on reptilesmagazine.com August 23, 2018, accessed April 12, 2021.
3. This means that evolutionary ideas of kinds turning into different kinds fly in the face of both the Bible and biology.

Acadia: The Maine Jewel Among America's Parks

1. Marsi, R. 1996. Acadia. In *National Parks*. C. Jackson, ed. White Plains, NY: Readers Digest Association, 8-17. See also Acadia National Park: Crown Jewel of the North Atlantic Coast. National Park Service. Posted on nps.gov, accessed April 22, 2021.
2. Sutton, A. and M. Sutton. 1985. *Eastern Forests*. New York: Chanticleer Press/Audubon Society Nature Guide, 139 (plate 6), 355-357, 360-366, 384-386, 398-399, 407-408, 417-419, 425-426; Johnson, J. J. S. 2012. Valuing God's Variety. *Acts & Facts*. 41 (9): 8-9.
3. Clarey, T. 2020. *Carved in Stone: Geological Evidence of the Worldwide Flood*. Dallas, TX: Institute for Creation Research, 194-233.
4. Rosen, J. 2016. Crystal clocks. *Science*. 354 (6314): 822-825; Ruprecht, P. and T. Plank. 2013. Feeding andesitic eruptions with a high-speed connection from the mantle. *Nature*. 500 (7460): 68-72. See Phelps, P. R., C.-T. A. Lee, and D. M. Morton. 2020. Episodes of fast crystal growth in pegmatites. *Nature Communications*. 11: 4986.
5. Niering, W. A. 1998. *Wetlands*. New York: Chanticleer Press, 147 (plate 28), 377-379, 388-393, 406-407, 499, 609.
6. Amos, W. H. and S. H. Amos. 1989. *Atlantic and Gulf Coasts*. New York: Chanticleer Press, 154-155 (plates 8-12), 504, 525, 534, 547, 605.
7. Johnson, J. J. S. 2013. Fulfilling the Genesis Mandate While Helping the Poor. *Acts & Facts*. 42 (12): 19. See also Johnson, J. J. S. 2013. Siberian Huskies and the Dominion Mandate. *Acts & Facts*. 42 (6): 18-19.
8. Acadia National Park: Invertebrates. National Park Service. Posted on nps.gov, accessed May 1, 2021. See Johnson, J. J. S. 2015. The Ghost Army. *Acts & Facts*. 44 (11): 20.
9. See Acadia National Park, ibid, note 8.
10. Johnson, J. J. S. 2018. Lobsters Get Comfortable in Their New Skin. *Acts & Facts*. 47 (9): 21.
11. Johnson, J. J. S. Common Seals Display Extraordinary Bioengineering. *Creation Science Update*. Posted on ICR.org May 5, 2020, accessed May 1, 2021.
12. Sherwin, F. Brittle Stars See with Their Skin. *Creation Science Update*. Posted on ICR.org December 26, 2019, accessed May 1, 2021.
13. Acadia National Park: Moths of Acadia. National Park Service. Posted on nps.gov, accessed April 22, 2021.
14. Hatler, D. F., D. W. Nagorsen, and A. M. Beal. 2008. *Carnivores of British Columbia*. Victoria, BC: Royal British Columbia Museum, 220-221.
15. Acadia National Park: Invertebrates: Winter. National Park Service. Posted on nps.gov, accessed April 22, 2021.
16. Psalm 150:6; Revelation 4:11.

Crater of Diamonds State Park and the Origin of Diamonds

1. Clarey, T. 2020. *Carved in Stone: Geological Evidence of the Worldwide Flood*. Dallas, TX: Institute for Creation Research, 285.
2. RATE stands for Radioisotopes and the Age of the Earth. The multifaceted project's predictions were published in 2000, and their spectacular fulfillments were published in 2005. All documents are available online at ICR.org/rate.
3. Given the relatively short half-life of 5,730 years for radiocarbon's decay into nitrogen, none would remain after 100,000 years.
4. Baumgardner, J. 2005. Carbon-14 Evidence for a Recent Global Flood and a Young Earth. In *Radioisotopes and the Age of the Earth: Results of a Young-Earth Creationist Research Initiative*. L. Vardiman et al, eds. San Diego, CA: Institute for Creation Research and Chino Valley, AZ: Creation Research Society, 587-630.
5. Taylor, R. E. and J. R. Southon. 2007. Use of natural diamonds to monitor ^{14}C AMS instrument backgrounds. *Nuclear Instruments and Methods in Physics Research B*. 259: 282-287.
6. Taylor, R. E., J. R. Southon, and G. M. Santos. 2018. Misunderstandings Concerning the Significance of AMS Background ^{14}C Measurements. *Radiocarbon*. 60 (3): 727-749.
7. Cupps, V. R. and B. Thomas. 2019. Deep Time Philosophy Impacts Radiocarbon Measurements. *Creation Research Society Quarterly*. 55 (4): 212-222.

Waco Mammoth National Monument

1. Hebert, J. 2018. The Bible Best Explains the Ice Age. *Acts & Facts*. 47 (11): 10-13.
2. Vardiman, L. 2003. Hypercanes Following the Genesis Flood. In *Proceedings of the Fifth International Conference on Creationism*. R. L. Ivey Jr., ed. Pittsburgh, PA: Creation Science Fellowship: 17–28.
3. *Waco Mammoth Site Special Resource Study / Environmental Assessment*. 2008. National Park Service.
4. Discordant "ages" are the norm in archaeology as well as geology. See for example, Thomas, B. Cherry-Picked Age for Key Evolutionary Fossil. *Creation Science Update*. Posted on ICR.org June 8, 2017, accessed July 13, 2021.; Thomas, B. 2020. How Not to Date Hominid Remains. *Acts & Facts*. 49 (6): 15; Austin, S. A., and A. A. Snelling. 1998. Discordant Potassium-Argon Model and Isochron "Ages" for Cardenas Basalt (Middle Proterozoic) and Associated Diabase of Eastern Grand Canyon, Arizona. In *Proceedings of the Fourth International Conference on Creationism*. R. E. Walsh, ed. Pittsburgh, PA: Creation Science Fellowship: 35–51.
5. See references in Thomas, B. 2018. When Did Noah's Flood Happen? *Acts & Facts*. 47 (3): 20.

Carlsbad Caverns National Park: Fast Formations

1. Nelson, V. 2020. *Catastrophic Caves*. Red Deer, AB. Untold Secrets of Planet Earth Publishing Company.
2. Oard, M. J. 2006. *Frozen in Time*. Green Forest, AR: Master Books, 41-44, 83-85.
3. Collins, M. J. et al. 1995. A Basic Mathematical Simulation of the Chemical Degradation of Ancient Collagen. *Journal of Archaeological Science*. 22 (2): 175-183.

Haleakala National Park: One of Many Young-Looking Volcanoes

1. Jones, C. Hawaii's Kilauea volcano erupts, lava fountains form in park. Associated Press. Posted on APnews.com September 29, 2021, accessed October 1, 2021.
2. Hawaiian stick spiders re-evolve the same three guises every time they island hop. *Cell Press*. Posted on ScienceDaily.com March 8, 2018, accessed October 1, 2021.
3. Gillespie, R. G. et al. 2018. Repeated Diversification of Ecomorphs in Hawaiian Stick Spiders. *Current Biology*. 28: 941-947, referencing Brower, A. V. 1994. Rapid morphological radiation and convergence among races of the butterfly *Heliconius erato* inferred from patterns of mitochondrial DNA evolution. *Proceedings of the National Academy of Sciences*. 91 (14): 6491-6495.

4. Snelling, A. A. 1998. Andesite Flows at Mt. Ngauruhoe, New Zealand, and the Implications for Potassium-argon "Dating." In *Proceedings of the Fourth International Conference on Creationism*. Pittsburgh, PA: Creation Science Fellowship.
5. Dalrymple, G. B. 1969. $^{40}Ar/^{36}Ar$ Analyses of Historic Lava Flows. *Earth and Planetary Science Letters*. 6: 47-55.
6. Krummenacher, D. 1970. Isotopic Composition of Argon in Modern Surface Volcanic Rocks. *Earth and Planetary Science Letters*. 8: 109-117.
7. Noble, C. S. and J. J. Naughton. 1968. Deep-Ocean Basalts: Inert Gas Content and Uncertainties in Age Dating. *Science*. 162: 265-267.
8. Austin, S. A. et al. 1994. Catastrophic Plate Tectonics: A Global Flood Model of Earth History. In *Proceedings of the Third International Conference on Creationism*. R. E. Walsh, ed. Pittsburgh, PA: Creation Science Fellowship.
9. Fletcher, C. H. et al. National Assessment of Shoreline Change: Historical Shoreline Change in the Hawaiian Islands. U.S. Geological Survey Open-File Report, 2011-1051, 55.

Crater Lake National Park: Serene Beauty After Volcanic History

1. Hopson, R. F. 2018. Crater Lake National Park, Southwest Oregon. In *The Geology of National Parks*, 7th ed. D. Hacker, D. Foster, and A. G. Harris, eds. Dubuque, IA: Kendall-Hunt, 649-666.
2. Macy, M. 1999. Crater Lake. In *America's Spectacular National Parks*. L. B. O'Connor and D. Levy, eds. Los Angeles, CA: Perpetua Press, 120-121.
3. Clarey, T. 2020. *Carved in Stone: Geological Evidence of the Worldwide Flood*. Dallas, TX: Institute for Creation Research.
4. Clarey, T. Plate Subduction Beneath China Verifies Rapid Subduction. *Creation Science Update*. Posted on ICR.org December 23, 2020, accessed October 29, 2021.
5. Clarey, T. 2019. Subduction Was Essential for the Ice Age. *Acts & Facts*. 48 (3): 9.
6. Cinder cones are made of a rock called scoria. These are volcanic rocks that contain many holes from gas bubbles. Cinders are about the same size as lava rocks for landscaping and/or gas grills, around 0.08–2.5 inches in length.
7. Birds. Crater Lake National Park, Oregon. National Park Service. Posted on nps.gov, accessed November 1, 2021. Regarding animals in parks and wildlife refuges of the Great West, see also Johnson, J. J. S. Yes, Deer, It's Time for Some Calm News. *Creation Science Update*. Posted on ICR.org April 27, 2020, accessed October 29, 2021. This article's coauthor visited Crater Lake and Lava Butte, Oregon, during the summer of 1992 for biogeography research for use in teaching ecology and ornithology at Dallas Christian College.

Pictured Rocks National Lakeshore: The Blanket Sand from Noah's Flood

1. Blewett, W. L. 2012. *Geology and Landscape of Michigan's Pictured Rocks National Lakeshore and Vicinity*. Detroit, MI: Wayne State University Press.
2. Clarey, T. 2020. *Carved in Stone: Geological Evidence of the Worldwide Flood*. Dallas, TX: Institute for Creation Research, 194-215.
3. Clarey, T. 2017. Resolving the Sheet Sand Enigma. *Acts & Facts*. 46 (11): 14.
4. Genesis 7:11. See also Clarey, *Carved in Stone*, 172-193.
5. Blewett, *Geology and Landscape of Michigan's Pictured Rocks*, 100.

Yellowstone National Park, Part 1: A Flood Supervolcano

1. Tweit, S. J. 1999. Yellowstone. In *America's Spectacular National Parks*. L. B. O'Connor and D. Levy, eds. Los Angeles, CA: Perpetua Press, 76-79.
2. Austin, S. A. 1998. The Declining Power of Post-Flood Volcanoes. *Acts & Facts*. 27 (8).
3. A hot spot is thought to form from a near-stationary deep mantle plume of high heat capable of producing a melt through the overriding crust.
4. Clarey, T. Plate Subduction Beneath China Verifies Rapid Subduction. *Creation Science Update*. Posted on ICR.org December 23, 2020, accessed February 12, 2022; Clarey, T. 2020. *Carved in Stone: Geological Evidence of the Worldwide Flood*. Dallas, TX: Institute for Creation Research.
5. How big is the magma chamber under Yellowstone? U.S. Geological Survey FAQ. Posted on usgs.gov, accessed February 17, 2022.
6. Hacker, D. and D. Foster, 2018. Yellowstone National Park: Northwest Wyoming, Eastern Idaho, Southern Montana. In *The Geology of National Parks*, 7th ed. D. Hacker, D. Foster, and A. G. Harris, eds. Dubuque, IA: Kendall-Hunt, 765-791.
7. Milstein, M. Old Faithful slows, but grows. *Billings Gazette*. First captured online January 17, 2001. Retrieved from web.archive.org January 11, 2022.
8. Theurer, J. Colors of Curiosity: Yellowstone's Microbes. *Yellowstone Quarterly*. Posted on yellowstone.org April 16, 2019, accessed January 13, 2022.
9. Geiling, N. The Science Behind Yellowstone's Rainbow Hot Spring. *Smithsonian Magazine*. Posted on smithsonianmag.com May 7, 2014, updated May 12, 2016, accessed January 14, 2022.

Yellowstone National Park, Part 2: Canyons and Catastrophe

1. Tweit, S. J. 1999. Yellowstone. In *America's Spectacular National Parks*. L. B. O'Connor and D. Levy, eds. Los Angeles, CA: Perpetua Press, 76-79.
2. Clarey, T. 2017. Minuscule Erosion Points to Hawaii's Youth. *Acts & Facts*. 46 (1): 9.
3. Hacker, D. and D. Foster. 2018. Yellowstone National Park: Northwest Wyoming, Eastern Idaho, Southern Montana. In *The Geology of National Parks*, 7th ed. D. Hacker, D. Foster, and A. G. Harris, eds. Dubuque, IA: Kendall-Hunt, 765-791.
4. Morris, J. 2011. The Channeled Scablands. *Acts & Facts*. 40 (10): 15.
5. Hebert, J. 2018. The Bible Best Explains the Ice Age. *Acts & Facts*. 47 (11): 10-13.
6. Austin, S. A. 1986. Mt. St. Helens and Catastrophism. *Acts & Facts*. 15 (7).
7. Thomas, B. and T. Clarey. 2021. The Painted Desert: Fossils in Flooded Mud Flats. *Acts & Facts*. 50 (4): 16-19.

Grand Teton National Park: The Youngest Rocky Mountains

1. The area of Grand Teton National Park was first surveyed by geologist James Stevenson in 1872 as part of the Ferdinand Hayden expedition to Yellowstone. See Clarey, T. and B. Thomas. 2022. Yellowstone National Park, Part 1: A Flood Supervolcano. *Acts & Facts*. 51 (3): 10-13.
2. Červený, J. et al. 2011. Directional preference may enhance hunting accuracy in foraging foxes. *Biology Letters*. 7 (3): 355-357. See also Krulwich, R. 'You're Invisible, But I'll Eat You Anyway.' Secrets Of Snow-Diving Foxes. NPR. Posted on npr.org January 3, 2014, accessed January 14, 2014.

3. Darwin, C. 1859. *On the Origin of Species*, 1st ed. Project Gutenberg eBook no. 1228.
4. Trut, L., I. Oskina, and A. Kharlamova. 2009. Animal evolution during domestication: the domesticated fox as a model. *Bioessays*. 31 (3): 349-360.
5. Hacker, D. and D. Foster. 2018. Grand Teton National Park. In *The Geology of National Parks*, 7th ed. D. Hacker, D. Foster, and A. G. Harris, eds. Dubuque, IA: Kendall-Hunt.
6. Clarey, T. 2019. European Stratigraphy Supports a Global Flood. *Acts & Facts*. 48 (12): 10-12.
7. Most of the Cenozoic system corresponds to the Tejas Megasequence, which refers to a vast stack of coarse-to-fine-grained layers deposited during the receding phase of the Flood. The Flood deposited six megasequences, and every continent carries some of each one.
8. The Teton fault is a large normal fault. These develop where one side of the fault drops and the other side moves upward, usually during an episode of crustal extension. The Teton fault has about 30,000 feet of offset.
9. Clarey, T. 2020. *Carved in Stone: Geological Evidence of the Worldwide Flood*. Dallas, TX: Institute for Creation Research, 114-151.
10. Hebert, J. 2018. The Bible Best Explains the Ice Age. *Acts & Facts*. 47 (11): 10-13.
11. Guliuzza, R. J. 2019. Engineered Adaptability: Continuous Environmental Tracking Wrap-Up. *Acts & Facts*. 48 (8): 17-19.

Mackinac Island State Park: Flood, Ice Age, and...Digestion?

1. Milstein, R. L. 1987. Mackinac Island State Park, Michigan. In *North-Central Section of the Geological Society of America: Centennial Field Guide*, vol. 3. D. L. Biggs, ed. Boulder, CO: Geological Society of America, 285-288.
2. Clarey, T. 2020. *Carved in Stone: Geological Evidence of the Worldwide Flood*. Dallas, TX: Institute for Creation Research.
3. Porter, P. 2013. *Historic Mackinac Island Visitor's Guide: Tours, History, Maps*. Mackinac Island, MI: Mackinac Island State Park Commission.
4. Clarey, T. 2020. Lava Flows Disqualify Lake Spillover Canyon Theory. *Acts & Facts*. 49 (10): 10-12.
5. Oard, M. J. 2004. *Frozen in Time: The Wooly Mammoth, The Ice Age, and the Bible*. Green Forest, AR: Master Books.
6. Widder K. R. 2006. *Dr. William Beaumont: The Mackinac Years*. Mackinac Island, MI: Mackinac Island State Park Commission.
7. Beaumont, W. 1838. *Experiments and Observations on the Gastric Juice and the Physiology of Digestion*. Edinburgh: MacLachlan and Stewart.

Appendix: Mount St. Helens, A Living Laboratory

1. Wicander, R. and J. S. Monroe. 2016. *Historical Geology: Evolution of Earth and Life Through Time*, 8th ed. Boston, MA: Cengage Learning.
2. Austin, S. A. 1986. Mt. St. Helens and Catastrophism. *Acts & Facts*. 15 (7).
3. Schieber, J., J. Southard, and K. Thaisen. 2007. Accretion of Mudstone Beds from Migrating Floccule Ripples. *Science*. 318 (5857): 1760-1763.
4. Snelling, A. A. 2009. *Earth's Catastrophic Past*, vol. 2. Dallas, TX: Institute for Creation Research, 493-499.
5. Boggs Jr., S. 2006. *Principles of Sedimentology and Stratigraphy*, 4th ed. Upper Saddle River, NJ: Pearson/Prentice Hall, 159-167.
6. Schieber, J. et al. 2013. Experimental Deposition of Carbonate Mud from Moving Suspensions: Importance of Flocculation and Implications For Modern and Ancient Carbonate Deposition. *Journal of Sedimentary Research*. 83 (11): 1025-1031.
7. Blatt, H., G. Middleton, and R. Murray. 1980. *Origin of Sedimentary Rocks*, 2nd ed. Englewood Cliffs, NJ: Prentice-Hall, Inc.
8. Portenga, E. W. and R. R. Bierman. 2011. Understanding Earth's eroding surface with ^{10}Be. *GSA Today*. 21 (8): 4-10.
9. Morris, J. D. and S. A. Austin. 2003. *Footprints in the Ash: The Explosive Story of Mount St. Helens*. Green Forest, AR: Master Books.
10. Thorarinsson, S. 1967. *Surtsey: The New Island in the North Atlantic*. S. Eysteinsson, trans. New York: Viking Press.
11. An Island Laboratory. 2008. *Science Illustrated*. May/June, 42-47.
12. Tonra, C. M. et al. 2015. The rapid return of marine-derived nutrients to a freshwater food web following dam removal. *Biological Conservation*. 192: 130.
13. River ecosystems show 'incredible' recovery after dam removal. Ohio State News. Posted on new.osu.edu December 28, 2015, accessed March 5, 2020.
14. Mount St. Helens Recovery Slowed By Caterpillar. University of Maryland, College Park. *ScienceDaily*. Posted on sciencedaily.com November 16, 2005, accessed March 5, 2020.
15. Thompson, A. Mount St. Helens Still Recovering 30 Years Later. *Live Science*. Posted on livescience.com May 17, 2010, accessed March 5, 2020.

IMAGE CREDITS

Adobe Stock: 12b, 13bl
Scott Arledge: 13br
Bigstock Photo: 16-19, 24-25, 28, 38-42, 44, 46, 54-55, 58, 61, 64-65, 66t, 67b, 74-85, 91t, 94-95, 97, 104, 106, 109, 112, 115-119, 122t, 130-131, 132b, 133-135, 138-141, 145, 158, 167, 172, 175
Matt Bowles: 59
Erin Clarey: 174
Tim Clarey: 136, 176, 178, 183, 186, 189
Werner J. Davis: 33, 92
Edward Downing : 20-21
Dlthewave via Wikipedia: 160l
Dreamstime: 136-137
Fanboyphilosopher (Neil Pezzoni): 47
Florida Department of Environmental Protection: 108
Eric Gaba via Wikipedia: 181
Google maps: 51r, 93l, 105, 191
Michael Hansen: 149, 161
iStockPhoto: 66b, 100, 110-111
Joel Kautt: 146-148, 150-152, 163-164, 168
Reinhard Link: 107
Allyssa Loveness: 121
Clint Loveness: 122b, 125-126, 157, 160tm, tr, 167
John Morris: 185
Moxfyre via Wikipedia: 192-193
NASA: 37, 50-51, 62-63, 72, 86, 142, 154, 170, 180l, 190
National Park Service (NPS): 29, 36, 43, 49, 51l, 60, 62, 67t, 69-71, 73, 87, 114r, br, 118, 124, 143, 155, 171
Hal333 via Wikipedia: 88-89
Don Patton: 114l
Henry G. Peabody Photographs, Greater Arizona Collection, Arizona State University Library, Tempe, AZ: 10-11
Public domain: 68, 91b, 177, 187t
George Roderick/University of California-Berkeley: 123
Daryl Robbins: 30-32
Shutterstock: 128-129
Springer Nature: *Nature Geoscience*: 26
James St. John via Wikipedia: 96
Brian Thomas: 22-23, 27, 34, 52-53, 56b, 57, 90, 98-99, 101, 103
Michele Thomas: 93r
Kevin Turley: 12t
USGS: 14-15, 63, 132t, 180r, 184, 187b
Rob Wegner: 45
Susan Windsor: 8-9, 56t, 102
Dave Woetzel: 113
Zion National Park: 57
Brenda Zaun, U.S. Fish & Wildlife: 121r

INDEX

This book was adapted from the following materials

Clarey, T. and B. Thomas. 2021. Grand Canyon: Exposing the Flood. *Acts & Facts*. 50 (1): 14-17.

Thomas, B. and T. Clarey. 2021. Arches National Park: Sculptures from the Flood. *Acts & Facts*. 50 (2): 16-19.

Thomas, B. and T. Clarey. 2021. Gunnison's Black Canyon: The Flood Solves Mysterious Missing Time. *Acts & Facts*. 50 (3): 16-19.

Thomas, B. and T. Clarey. 2021. The Painted Desert: Fossils in Flooded Mud Flats. *Acts & Facts*. 50 (4): 16-19.

Thomas, B. and T. Clarey. 2021. Zion National Park: Evidence of Deep Water Sand Waves. *Acts & Facts*. 50 (5): 16-19.

Thomas, B. and G. Parker. 2021. The Everglades: Two Biology Basics the Bible Got Right. *Acts & Facts*. 50 (6): 14-17.

Johnson, J. J. S. and T. Clarey. 2021. Acadia: The Maine Jewel Among America's Parks. *Acts & Facts*. 50 (7): 14-17.

Thomas, B. 2021. Crater of Diamonds State Park and the Origin of Diamonds. *Acts & Facts*. 50 (8): 14-17.

Thomas, B. 2021. Waco Mammoth National Monument. *Acts & Facts*. 50 (9): 14-17.

Parker, G. 2021. John Pennekamp Coral Reef State Park. *Acts & Facts*. 50 (10): 12-15.

Woetzel, D. and B. Thomas. 2021. Carlsbad Caverns National Park: Fast Formations. *Acts & Facts*. 50 (11): 14-17.

Thomas, B. 2021. Haleakala National Park: One of Many Young-Looking Volcanoes. *Acts & Facts*. 50 (12): 14-17.

Clarey, T. and J. J. S. Johnson. 2022. Crater Lake National Park: Serene Beauty After Volcanic History. *Acts & Facts*. 51 (1): 14-17.

Clarey, T. 2022. Pictured Rocks National Lakeshore: The Blanket Sand from Noah's Flood. *Acts & Facts*. 51 (2): 10-13.

Clarey, T. and B. Thomas. 2022. Yellowstone National Park, Part 1: A Flood Supervolcano. *Acts & Facts*. 51 (3): 10-13.

Clarey, T. and B. Thomas. 2022. Yellowstone National Park, Part 2: Canyons and Catastrophe. *Acts & Facts*. 51 (4): 10-12.

Clarey, T. and B. Thomas. 2022. Grand Teton National Park: The Youngest Rocky Mountains. *Acts & Facts*. 51 (5): 10-13.

Clarey, T. 2022. Mackinac Island State Park: Flood, Ice Age, and...Digestion? *Acts & Facts*. 51 (6): 10-13.

Clarey, T. and F. Sherwin. 2020. Mount St. Helens, Living Laboratory for 40 Years. *Acts & Facts*. 49 (5): 10-13.

Carved in Stone

Geological Evidence of the Worldwide Flood

Dr. Timothy Clarey

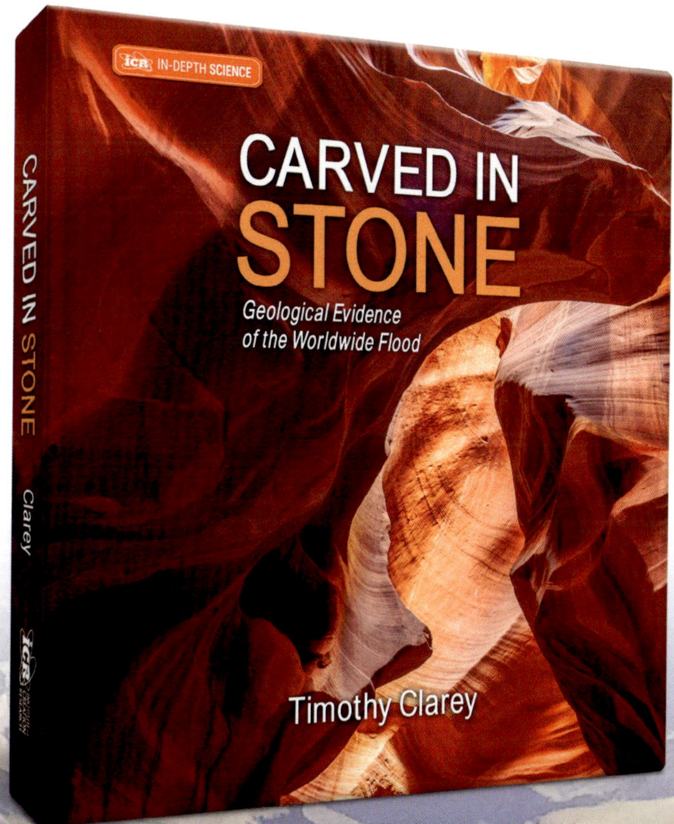

Genesis records a worldwide cataclysm that reshaped Earth's surface and destroyed almost all land-dwelling creatures. Conventional geologists insist the global Flood is a myth—but they've never studied the rock record across multiple continents simultaneously.

ICR geologist Dr. Timothy Clarey does just this in *Carved in Stone: Geological Evidence of the Worldwide Flood*. He examines the sedimentary rock record continent by continent, layer by layer, and finds that the data provide clear evidence of a year-long progressive flood just as described in the Bible. The rocks do not lie!

Find out more about this book and other resources at **ICR.org/store**

About the Institute for Creation Research

At the Institute for Creation Research, we want you to know God's Word can be trusted with everything it speaks about—from how and why we were made, to how the universe was formed, to how we can know God and receive all He has planned for us.

That's why ICR scientists have spent more than 50 years researching scientific evidence that refutes evolutionary philosophy and confirms the Bible's account of a recent and special creation. We regularly receive testimonies from around the world about how ICR's cutting-edge work has impacted thousands of people with God's creation truth.

How Can ICR Help You?

You'll find faith-building science articles in *Acts & Facts*, our bimonthly science news magazine, and spiritual insight and encouragement from *Days of Praise*, our quarterly devotional booklet. Sign up for FREE at **ICR.org/subscriptions**.

Our radio programs, podcasts, online videos, and wide range of social media offerings will keep you up to date on the latest creation news and announcements. Get connected at **ICR.org**.

We offer creation science books, DVDs, and other resources for every age and stage at **ICR.org/store**.

Learn how you can attend or host a biblical creation event at **ICR.org/events**.

Discover how science confirms the Bible at our Dallas museum, the ICR Discovery Center. Plan your visit at **ICRdiscoverycenter.org**.

ICR INSTITUTE FOR CREATION RESEARCH

P. O. Box 59029
Dallas, TX 75229
800.337.0375
ICR.org